The Last Wild Wolves

IAN McALLISTER

With Contributions by Chris Darimont

Introduction by Paul C. Paquet

THE LAST WILD WOLVES

Ghosts of the Rain Forest

University of California Press

BERKELEY   LOS ANGELES

For Karen, Callum, and Lucy,
who make for the best of all journeys.

University of California Press, one of the most distinguished university presses in the United States, enriches lives around the world by advancing scholarship in the humanities, social sciences, and natural sciences. Its activities are supported by the UC Press Foundation and by philanthropic contributions from individuals and institutions. For more information, visit www.ucpress.edu.

University of California Press
Berkeley and Los Angeles, California

Published by arrangement with Greystone Books
A division of Douglas & McIntyre Ltd.
2323 Quebec Street, Suite 201
Vancouver, British Columbia
Canada V5T 4S7
www.greystonebooks.com

Library of Congress Cataloging-in-Publication Data

McAllister, Ian, 1969–.
  The last wild wolves : ghosts of the rain forest / Ian McAllister ; with an introduction by Paul C. Paquet and contributions by Chris Darimont.
    p.   cm.
  Simultaneously published in Vancouver by Greystone Books.
  Includes bibliographical references and index.
  ISBN 978-0-520-25473-2 (cloth : alk. paper)
  1. Wolves—British Columbia.   I. Title.

QL737.C22M374   2007
599.77309711—dc22   2007010887

Editing by Nancy Flight
Copy editing by Wendy Fitzgibbons
Jacket and text design by Naomi MacDougall
Jacket photographs by Ian McAllister
Map on page 9 by Stuart Daniel/Starshell Maps
Printed on paper that comes from sustainable forests managed under the Forest Stewardship Council
Manufactured in China

16   15   14   13   12   11   10   09   08   07
10   9   8   7   6   5   4   3   2   1

Contents

Introduction

Paul C. Paquet, PhD

SENIOR SCIENTIST, *Raincoast Conservation Foundation*

IN THE PAST, the elusive gray wolf roamed most of the Northern Hemisphere, including nearly all of Eurasia and North America. Except for humans and possibly the African lion, gray wolves once had the most extensive range of any terrestrial mammal. They were found in a variety of environments, from dense forest to open grassland and from the Arctic tundra to extreme desert, avoiding only swamps and tropical rain forests. Now wolves occur mostly in remote and undeveloped areas of the world with sparse human populations. These few remaining secluded areas of relatively unmodified landscapes are biological and cultural treasures, our last opportunities to preserve the highly specialized and co-evolved relationships that are being replaced elsewhere with invasive species and managed landscapes.

On the North American mainland, gray wolves were once found everywhere except the southeastern United States, California west of the Sierra Nevada range, and the tropical and subtropical regions of Mexico. The species also occurred on large continental islands, such as Newfoundland and Vancouver Island, on smaller islands off coastal British Columbia and southeast Alaska, and throughout the Arctic Archipelago and Greenland but was absent from Prince Edward Island, Anticosti, and Haida Gwaii (the Queen Charlotte Islands).

Over the past four centuries, gray wolf populations in North America have been decimated by overwhelming increases in the human population, development of agriculture, and expansion of industrial forestry. Moreover, hundreds of thousands of wolves have been trapped, poisoned, shot from helicopters, and sterilized, among other modes of destruction. By the beginning of the twentieth century, wolves had nearly vanished from the eastern United States, most of southern Canada, and the Canadian Maritime provinces. By 1960, the wolf had been exterminated by federal and state governments from all of the United States except Alaska and northern Minnesota. The distribution of the gray wolf in North America is now confined primarily to Alaska and Canada. In the conterminous United States, populations in northern Minnesota, northern Wisconsin, and Michigan's Upper Peninsula, as well as parts of Arizona, Washington, Idaho, Montana, and Wyoming, have been augmented by wolves introduced from Canada and by natural recolonization.

Canada's North Pacific coast supports the largest intact temperate rain forest left on the planet.

In Canada the gray wolf is still found throughout much of its historical range, including coastal islands. Cessation of most wolf control programs has allowed the wolf to recover in many areas where it had been extirpated, although the species is completely gone from Newfoundland, Nova Scotia, and New Brunswick and is absent or rare in the densely populated and developed parts of the other provinces. In many areas within the wolf's worldwide range, populations have been decimated or completely extirpated, making Canada an important stronghold of the species.

WHERE THE LAND meets the sea on British Columbia's untamed Pacific Coast, a distinctive subspecies of gray wolf lives as it always has, comparatively unaffected by people. The Pacific Ocean overwhelmingly defines and influences the wolves' environment, which is rich in human culture and natural history. Distributed on the mainland and nearby islands, these wolves swim in open ocean between landmasses as distant as 13 kilometres, contending with erratic winds, cold water, and strong tidal currents. Their physical features, behaviours, and pack traditions have been shaped by millennia of adaptation to the region's marine-dominated rain forest. Here wolves intermingle with other wide-ranging species, such as grizzly bears, killer whales, humpback whales, salmon, and migratory birds, many of which have been exterminated from much of their former ranges.

Among all the regions of North America where wolves still roam, the central and north coasts of British Columbia and the nearby offshore islands are ecologically unique. A network of islands, waterways, and mountains naturally fragments the land- and seascape. Yet the marine and the terrestrial are inextricably linked, providing the food and shelter that sustain coastal wolves. This remote ocean archipelago comprises North America's most unusual and one of its most pristine wolf populations.

Many coastal wolves are island dwellers whose territories include groups of islands. Consequently, they are compelled to travel on land and swim between distant landmasses to make a living. Many of their prey, as well as other carnivores, such as black bears and grizzly bears, do the same. Although water barriers may constrain the movements of wolves and their prey, the ocean also augments the food available on land. In addition to feeding on deer, moose, and goat, wolves eat salmon, clams, crabs, and marine carrion such as beached seals, whales, and squid. In the fall, spawning salmon constitute a considerable part of the wolf diet as the fish return to the rivers and creeks of the rain forest, which are also used by wolves, bears, and other terrestrial species to travel among estuaries and to reach inland forests. Like bears, wolves transport marine nutrients from waterways into the region's ancient forests. Abandoned salmon carcasses, wolf feces, and wolf urine feed a diversity of organisms and fertilize coastal ecosystems.

What has become widely known as the Great Bear Rainforest of coastal British Columbia presents one of the last opportunities on Earth to conserve wildlands that still support native species, maintain unfettered ecological and evolutionary processes, and contain rare ecosystems such as old-growth forests in their natural state. But large-scale clear-cut logging on the mainland and the more remote islands of the central and north coast is threatening the future of the Great Bear Rainforest. In addition, threats to the marine environment by overfishing, oil and gas exploration, and fish farming continue to escalate.

Apart from living in a secluded and relatively intact region, what makes the coastal gray wolves of British Columbia so distinctive, and why do biologists and other people throughout the world view them with such fascination, curiosity, and concern? Largely, it is because our familiarity with the "typical" behaviour and ecology of wolves—an awareness augmented by our intimate knowledge of domestic dogs and a recent spate of scientific studies, TV documentaries, magazine articles, and books—is challenged by the captivating incongruity of coastal wolves. These wolves do not behave the way we expect wolves to behave. Moreover, attitudes towards gray wolves have changed drastically in recent years. Now conservation groups and large sectors of the public regard wolves positively and believe that protecting and preserving them is a high priority.

Many are also drawn to wolves because they are iconic emblems of wild nature. We understand that our expanding and all-consuming civilization has led to the demise of wolves, and we are concerned about further human encroachment into their wild sanctuaries. In this human-

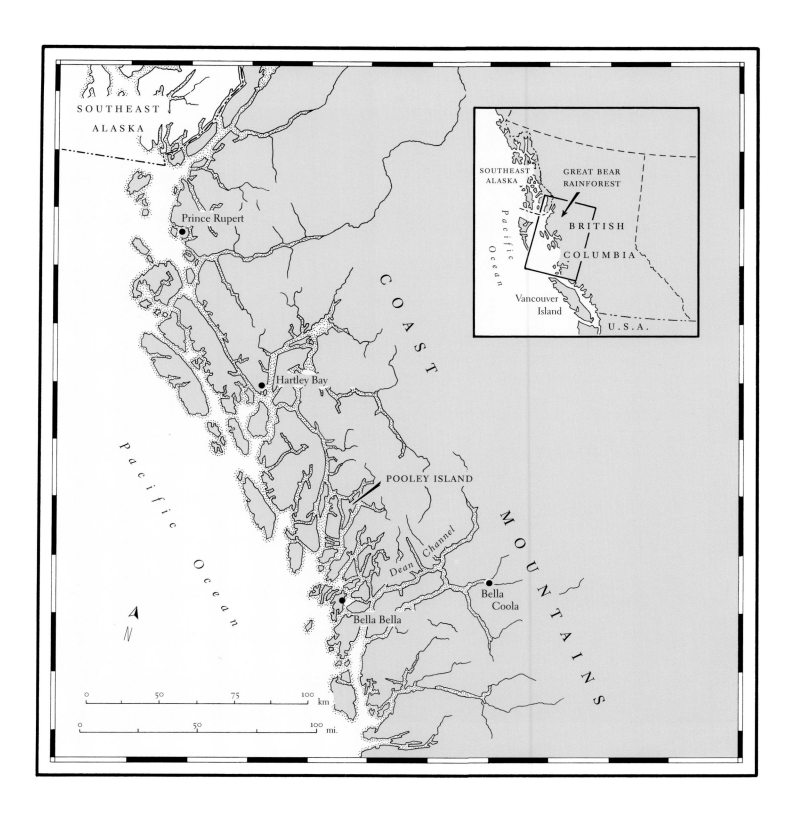

SOUTHEAST
ALASKA

Prince Rupert

COAST

Hartley Bay

POOLEY ISLAND

Pacific Ocean

Dean Channel

MOUNTAINS

Bella
Coola

Bella Bella

N

0   50   75   100 km

0   50   100 mi.

SOUTHEAST
ALASKA

GREAT BEAR
RAINFOREST

BRITISH

COLUMBIA

Pacific
Ocean

Vancouver
Island

U.S.A.

Introduction

dominated world, the conditions that wolves require to survive are quickly becoming rare commodities. People are rightly worried that wolves everywhere are vulnerable to disturbance. Even the largest Canadian and U.S. parks and reserves are inadequate in area to protect wolves fully.

Clearly, as wolves disappear, so does much of the wild nature that wolves signify and that people depend on for spiritual nourishment and physical sustenance. Wolves and people are both understood to be victims of unrestrained industrial progress. From this perspective, wild wolves living in wild areas of the Great Bear Rainforest provide hope for many that not all has been lost. Ironically, the species once regarded as a threat to our survival is turning out to be a test of how likely we are to live sustainably in the natural world.

In 1998 I met Ian McAllister at a meeting in Victoria, B.C. Ian was already well known for his wildlife photography and his unrelenting advocacy on behalf of coastal grizzly bears and ancient forests. From his vantage point in Bella Bella, where he and his wife, Karen, still live, these were his rain forest neighbours in need of his support. As a co-founder of the Raincoast Conservation Society, Ian coined the term "Great Bear Rainforest" as a moniker

for the central coast. I was impressed and moved by Ian's knowledge of natural history, coupled with his unapologetic zeal and enthusiasm for protecting the pristine environment of coastal British Columbia from degradation. However, his stories of island-hopping, salmon-eating wolves were particularly striking. We both realized that given the imminent environmental threats of industrial forestry, the lives of these remarkable coastal wolves needed to be documented and their status determined by serious scientific study. Accordingly, we began a search for someone with academic qualifications who could work closely with Raincoast and First Nations on such a study.

Some nine months later, at a celebration of my fiftieth birthday, Chris Darimont introduced himself and expressed his interest in working with the coastal wolves. At the time, he was involved with a wolf research project that I had initiated ten years earlier in the Rocky Mountains of Banff National Park. From the first moments of that initial meeting, I knew that Chris was the right person to lead the fieldwork. He was smart, experienced in the field, inquisitive, and genuinely respectful of others. He listened more than he spoke, and when he did speak, he usually asked questions. He was the perfect candidate, a young and unpretentious scientist who was more than willing to learn from First Nations.

Prospects for a serious study of coastal wolves improved considerably when Chris was accepted as a graduate student in biology at the University of Victoria under the supervision of Dr. Tom Reimchen, an exceptional researcher I had long admired and still consider one of Canada's brightest and most inventive ecologists. Dr. Reimchen became the scholarly and intellectual guide for the research. The addition of Chester Starr from the Heiltsuk First Nation to the study team provided the long-established wisdom of local people and an important guide, counsellor, and mentor for Chris. In this book the expressive lens and narrative of Ian McAllister tells the story of how Chris and Chester together discovered the heart and spirit of the coastal wolves.

The story leaves little doubt that if we hope to grasp the essence of this unusual wolf, we are compelled to know and understand the complex landscape and seascape that over millennia have shaped it. Through Ian's images and words, we are placed in the very environment where coastal wolves hunt, play, and raise their families; we physically and emotionally experience that landscape. The true nature of these wolves is revealed to those willing to abandon inflexible notions and embrace the feelings that overwhelm all who encounter them in their distinctive rain forest setting. This emotional boundary is the margin where modern science and passion merge to provide a window into the natural world—a world we once inhabited and understood without thought.

The Last Wild Wolves is an accolade to all native species that depend on the Great Bear Rainforest for survival. Ian, Chris, and Chester eloquently speak for the overwhelming majority of species that are unable to speak for themselves. I am sure that given the opportunity, this silent and increasingly beleaguered majority would enthusiastically endorse the efforts being carried out on their behalf.

The interface between rain forest and ocean provides habitat for many species of wolf prey.

Apex Predator

. . .

IT WAS very near the end of the spawning season, and I was trying to squeeze in every last day of photography and observation
before the salmon were completely gone. The peaks surrounding Dean Channel on the central coast of British Columbia were
shining bright with a fresh layer of snow. The feet of my waders
were deep in the mud, covered in a mixture of decaying fish, rich
alluvial silt, fish scales, and bones. The gut-wrenching stench of
tens of thousands of spawned salmon permeated the valley. Spruce
needles and leathery flaps of salmon skin floated by in the tannin-
tinted waters. Maggots, submerged by the tide, rolled around like
rice kernels, devouring the grey slime that only weeks ago was a
silver, powerful salmon. I tried not to remember that I drink from
this river at other times of the year.

A five-year-old male grizzly
bear sniffs out a visitor.
Coastal grizzlies display
surprising tolerance of humans.

About 50 metres (160 feet) upriver, an old friend was busy sucking on the decaying corpses like an overgrown child surrounded by Häagen-Dazs ice cream. Only this diner, with white flesh smeared across his lower jaw, was a big old grizzly bear, mostly black, weighing close to three hundred kilograms (seven hundred pounds)—probably 25 per cent heavier than when I had first seen him in the spring. His belly was so distended that it dragged along the ground. I once heard that scientists analysing tissue samples from bears at this time of year find more traces of salmon than they do of the bears.

It was a lazy afternoon. I had counted a dozen grizzly bears here overall in the previous few weeks, including a mother with three of the year's cubs feeding near the lower river. I saw tracks of black bears but rarely saw the bears; they preferred to feed at night or at the less desirable fishing spots—away from the grizzlies. All had been feeding on salmon for close to four months, recently for almost twenty hours a day, and were very near the caloric Zen state that bears need to reach just before heading into the snowy mountains to hibernate for the winter.

I was feeling somewhat in a Zen state myself and sat down in the mud, leaning my head back against a rain-soaked cedar. As I was about to close my eyes, I suddenly saw the grizzly stiffen and stand up on his hind legs, dropping his headless salmon. His nostrils flared, and he made a loud woofing sound. I followed his gaze to the other side of the estuary.

As if appearing in a dream, a stream of wolves emerged from the forest edge. By the time I counted thirteen of them, they had covered a quarter of the distance across the estuary. With their heads and tails up and ears forward, they fanned out across the mud flat, moving quickly and purposefully towards the grizzly. There was no question of what their intent was.

And they were beautiful. The family was at its fullest, before winter, disease, old age, or an errant deer or mountain goat hoof killed some of its members and before next year's pups were born. The pack coordinated their movements, running with a rhythm, with discipline, confidence, and a touch of attitude. The adults took the lead, and the pups, with their disproportionate clown feet and oversized ears, more like teenagers now, held back slightly.

Moments later, the wolves splashed through the shallow water and broke into a full sprint, ravens and gulls spilling

Each morning, this pack's
adults and pups joined for bonding
and playtime, a ritual that helps
establish the social hierarchy that
structures wolf families.

out of their way. When they were sixty metres away, the grizzly dropped down on all fours and took off like a race-horse bursting out of the gate. The huge bear galloped across the river, his salmon-filled belly swinging.

By the time I realized that I was situated directly between the bear and the nearest stand of trees, exactly in his path as he headed for cover, it didn't matter: I couldn't have moved in time if I had wanted to. Putting my hands in front of my face, I felt mud spraying across me and smelled the bear's breath, foul from his weeks of consuming rotting fish. Ten metres past and behind me, he smashed headfirst into the trees and underbrush. Afterwards I found an alder, fifteen centimetres (six inches) in diameter, broken in half and shattered.

Less than a minute earlier, I had been almost asleep.

As the next collective 450 kilograms (1,000 pounds) of carnivore headed my way, I was relieved to see the wolves slow down, their job apparently done. Once my breathing started up again, I found what followed even more astounding.

The pack reassembled in the middle of the mud flat, and a large, dark alpha male, the leader, began howling. Within seconds, every other member joined in. The sound

was like a victorious battle cry, and it seemed to silence every living thing in the valley. Even the songbirds stopped their singing to listen. I think the wolves were just ensur-ing that the bear would not have any second thoughts about fleeing, but I doubt that he did.

Then, as quickly as the chase had started, the wolves began playing with each other. Subordinate pups lay on their backs while their dominant siblings jumped over and on top of them. The youngsters chewed on each other's ears and legs, as well as on driftwood, seaweed, and other treasures washed up on the estuary; ran in circles, playing a sort of wolf tag; peed, and scratched the ground. They tripped over their floppy feet. The adults vacillated between indifference and full attention, and all of this had meaning, as individuals—adults and pups—worked to sort out their places in the very social and hierarchical world of wolves. It was tough for me to keep track of all this tomfoolery. Even-tually the adults lay on the cool ground, barely panting, watching the pups at play.

It was as if the recent attack on a full-grown grizzly bear was just a typical event in a day in the life of this pack of rain forest wolves. With demonstrably little effort, they had

A "spirit" or Kermode bear cub stays close to its mother. Kermodes are a subspecies of black bear and may be black or white.

sent one of North America's largest and fiercest land mammals packing. The grizzly did not hesitate when he registered what was coming his way; this was not the first time such an onslaught had happened.

I was amazed, almost as much by the bold, premeditated attack as by the nonchalance of the wolves afterwards. Clearly, they knew their place here.

A few years earlier, in the early nineties, I had studied the coastal rain forest from the perspective of the grizzly bear. Now I suddenly realized that I had missed out on an entirely different world. But I could count on my fingers the number of times I had encountered wolves, and the length of those encounters was often measured in seconds. How could a more dedicated study of them even be possible?

WHEN I FIRST started exploring what has since become known as the Great Bear Rainforest, I rarely saw a wolf. Kilometre after kilometre, month after month as my time in the forest turned into years of exploration of its many river valleys and offshore islands, the rain forest wolves remained hidden. I had only an occasional glimpse late in the evening.

Once when my wife, Karen, and I were on a bluff overlooking an estuary, waiting for a bear—grizzly, black, or the white "spirit" or Kermode bear, a subspecies of the black bear—to show itself, a wolf trotted out. It stuck to the tall sedges in the low depressions carved by the river in flood, hidden but for its ears; we followed its movements only by the sway of grass and sedge. Then it disappeared into the forest.

More often than not, the wolves showed themselves in other ways—a track etched in the mud, a few scats here and there, the well-chewed, moss-covered bones of a Sitka black-tailed deer, and, most frequently and possibly most grand of all, a late-evening chorus of howls heard from the deck of our boat at a lonely anchorage. The sound echoed softly off the high granite walls of some slope or side hill, somewhere where the wolves hunted in the vast sea of verdant rain forest.

I knew that wolves were among the planet's most elusive animals and that they were capable of travelling great distances—more than 70 kilometres (43 miles) in a day, or more likely during the night; the longest known trek was 177 kilometres, but that was in flat country. I thought they

Prologue

were opportunistic hunters, always on the move, searching for prey, and therefore unpredictable. I certainly didn't
think they would tolerate a human being in close proximity
for long.

I believed that wolves moved randomly and that my
early encounters with them had been coincidental. And
since they had no discernible patterns, I thought they could
not be studied or observed without the use of invasive techniques such as radio and satellite telemetry—especially not
by a relatively slow and clumsy (and, from a wolf's perspective, excessively smelly) night-blind human.

But was it the wolves that were random? Or was it my
search pattern that was all wrong? In addition, this wasn't
the tundra, where you could sit on top of a mountain undetected and with a telescope to observe wolves clearly from
many kilometres distant. Visibility in the coastal rain forest
is often measured by the extent of your arm and the way
sound and wind carry over the water.

The more I travelled the extremes of the north coast
of British Columbia, from the windswept offshore islands
to the icefields of the Coast Mountains range, the more I
understood that wolves dominate the landscape in a way
that grizzly bears cannot. Grizzly bears provide arguably
one of the best portals into the salmon forests that charac-
terize the larger B.C. mainland watersheds. As an "umbrella"
species, they indicate functioning ecosystems, and by their
sheer size and spirit, they remain a top-level icon for rain
forest wildness.

As omnivores, they also are more adaptable to different
conditions than are wolves; grizzlies have developed more
plasticity, or options, for survival. As well, because they
sleep through the less productive winter, they are virtually
seasonal residents of the coast.

Wolves, in contrast, are dedicated carnivores; when prey
(meat) cannot be hunted (or scavenged), they die. And often
they are providing not just for an individual or a small pack,
but for an extended family. Nor do they sleep through lean
times. Since cougars are rare in or absent from many areas
of this coast, wolves are *the* apex predator here. They trump
all others.

Wolves are also different from grizzlies in how they
roam the coast. I have found wolf tracks high on ridge lines,

Two grizzly bear cubs of the year learn from Mom as she searches the flats for clams, crabs, mussels, barnacles, and other intertidal critters.

at eighteen hundred metres (six thousand feet), following in the footsteps of their mountain goat prey. I have found them at sea level on the outer coastal extremes, separated from the mainland by kilometres of open ocean. Wolf packs, with their deadly efficient, strategic, and co-operative hunting techniques, have penetrated every niche of the coastal rain forest in ways that the more solitary and omnivorous grizzly bears cannot.

From almost the beginning, in a relatively short period of time, grizzly bears allowed me close access to observe their world. It did not take long to figure out the best habitat in which to find them. Their needs are relatively well documented; there are literally rooms full of reports, films, documentaries, and books on coastal grizzly bear ecology.

By contrast, I had a tough time finding any scientific information about the status or ecology of the wolves of B.C.'s north coast, though there was much to read about other wolves, including those to the north in Alaska. The rain forest wolf remained unstudied and mysterious outside First Nations culture. Science offered little, museums had no information, and even industry, which had plans to "develop" the coast, could provide no data or biological studies.

Here was the largest intact temperate rain forest left on the planet, harbouring a species whose North American range has been reduced by 40 per cent and individual numbers by 80 per cent in little over three hundred years, and yet the wolf's status, ecology, and behaviour remained largely unknown. Coastal wolves in North America once ranged from Mexico to Alaska, but by the 1920s they had been extirpated south of the Great Bear Rainforest.

When I visited First Nations communities or attended potlatches, however, I noticed signs of wolf everywhere. Wolf plays a large role in the people's lives. Wolf masks, poles, and art are prominently displayed alongside cultural crests of other important animals such as the blackfish, grizzly, and raven. When I talked to elders or was fortunate enough to listen to their stories, I heard wolves being described not as indiscriminate killers, an undeserved reputation for which they were savagely killed by those newcomers who settled continental North America, but as providers and protectors. In the remote Native communities of coastal B.C., wolf society and culture are revered.

Families describe their relationship with wolves over millennia with pride and clearly consider themselves privileged

if they belong to the Wolf clan. People are elevated in stature when their family holds the Wolf crest. Wolf (*K'vsls* in the Heiltsuk language) is ever-present. Some nations, like the Heiltsuk, turn to members of the Wolf clan for support or direction in troubling times, such as in wars or famines of the past. The Nuxalk describe the wolf as having sacred powers. The common theme in the old stories is that the wolf was willing to help humans and frequently transformed humans into wolves.

These are people who have lived near and with wolves for thousands of years and are very comfortable with them. Yet in the rest of North America, wolves are too often viewed with such unwarranted hatred or fear that many people kill them—not just to rid the land of wolves but also to exact vengeance, torturing them to make them "pay" for being wolves. These perspectives are so utterly at odds with each other that it is difficult to imagine they are held by the same species, on the same planet and the same continent.

An animal cannot maintain such a layered reputation through so many centuries, encoded in myth, misconceptions, and lore, without possessing an immense spirit, mystery, and intelligence. Would these rain forest wolves, having hidden from the terror wrought upon their continental kin, open their world to me? These were the thoughts that perplexed and intrigued me.

IT IS NOW more than a decade since I watched those wolves chase that grizzly off into the trees. The seasons that followed have reshaped my understanding of wolves and the role they play in the temperate rain forest. I named that wolf family the Fish Trap Pack, and I have watched them successfully raise a new generation of pups every year since. They were among the first to introduce me to their culture, their society.

Although no research technique can replace direct observation, some critical questions about coastal wolf ecology remained as elusive as the wolves themselves. How are the wolves related genetically to the rest of the wolves in North America? They certainly look different from wolves that I had seen elsewhere, and they live in an environment that is unique on this planet. How many wolves lived in the rain forest? I wondered. How large was each pack's range? What was the extent of the wolves' diet? And, most important, what was necessary to protect them in the face of a rapidly changing coastline? Land use plans for the Great Bear Rainforest were being developed without taking into account the wolf.

When studying a species (such as bears) that sleeps throughout the winter, one can get away with a seasonal approach. But with wolves it was different. In 1998, Karen and I moved full-time to a house on Denny Island, across from the Native community of Waglisla (Bella Bella), in the heart of the Great Bear Rainforest, to expand on our work with the Raincoast Conservation Society, the wildlife conservation group we helped co-found in 1990.

Prologue

With the benefit of local knowledge, in particular from the Heiltsuk Nation, I was able to spend more focussed time with the wolves year-round. Our place quickly became a call centre for wolf sightings by locals and passing mariners. This community-based approach greatly increased our knowledge of wolves.

A chance encounter with wolf expert Paul Paquet in 1998 alerted me to the fact that these were the least studied wolves in all of North America. In particular, little was known about their genetics and feeding ecology. Paul introduced me to Chris Darimont, an undergraduate student at the University of Victoria, who had just finished a season volunteering on another wolf research project in the Rocky Mountains; his task had been to locate and follow radio-collared wolves from his truck. More often than not, he ended up finding them dead—shot, trapped, poisoned, or run over by cars or trains. Chris told me that he felt more like an undertaker than a wolf researcher.

I had learned by then that it was possible to follow the rain forest wolves and to observe them with minimal impact. But a more intensive study using traditional scientific methods meant capturing them, collaring them, and following them by plane or helicopter. These techniques were intrusive and answered only a limited set of questions. However, Chris and Paul assured me that recent advances in molecular research techniques, using only what the wolves leave behind—scat and hair—would provide the same answers and more. Although this data-collection method was more labour intensive for the researchers, the wolves would not be harmed or harassed in any way.

The Raincoast Conservation Society and local First Nations launched the Rainforest Wolf Project in 2000. The study area is huge—about 65,000 square kilometres (25,000 square miles) of the central and northern B.C. coast—and while information on wolves was gathered throughout this region, a subzone of some 3,000 square kilometres near Bella Bella was chosen to allow a more intensive and manageable core study area. This book reveals the groundbreaking findings from this research, as well as my own personal experiences with and observations of some of the wolf packs.

As I gained the confidence of the Fish Trap Pack and learned more about them, I became aware that another population of rain forest wolves lived in a very different

A grizzly bear greets the
Companion. My wife, Karen,
and son, Callum, watch
from the wheelhouse.

way and hidden far from people on the extreme outer coast. They made their living as much from the ocean as from the land, and I called them the Surf Pack. If I could gain the trust of these wolves, a more complete picture of coastal wolf ecology might emerge.

I must say a word about the photographs in this book. Not all packs have allowed me to observe and photograph them fishing as much as the Fish Traps have. The obtrusive process of entering and leaving a river to observe wolves fishing often provides enough disturbance to push them farther upriver, so sometimes I spend the night in tree platforms that I built earlier in the season over their favourite fishing spots. Because the wolves are often nocturnal, there is only a narrow window after sunrise and before sunset in which to view them. They are active at night and rest during the day. It is exciting but often frustrating to lie awake in my sleeping bag all night, listening to the splashes, the growls, the shattering sound of bones being crunched and gnawed, the playing and yelps; wolves can be noisy at night. Salmon after salmon is caught and consumed within sixty metres of my platform, yet when daylight slowly arrives and

I start to think of taking a picture, the pack drifts one by one back into the forest, where I won't see or hear from them again until that evening.

These nocturnal habits are why many people that I talk to, even those born and raised on the coast, have never seen a wolf. Just seeing a wolf in the rain forest is a gift, never mind photographing one.

The pictures in this book are the product of hundreds and hundreds of days and predawn starts that included many during which I never saw a wolf. For every day that I was able to capture an image on film, a week and sometimes several went by with no such luck.

Although this book draws from my own observations of more than forty packs observed over a seventeen-year period between Knight Inlet and the Alaskan panhandle, the following pages mostly describe the Fish Trap Pack and the Surf Pack. It is my hope that they will help serve as ambassadors for all the wolves of Canada's North Pacific coast.

Prologue

Wolves are mostly
nocturnal, but when out
and about during the
day they are adept at
remaining hidden by
their surroundings.

Stealth Season

SPRING

LAMA PASS, part of the Inside Passage, was cold and still, and my breath streamed behind me as I motored out of the bay. The sweet-smelling red-cedar smoke drifting from the stovepipes of Waglisla lay softly over the water. Only a few lights appeared in the village on this spring morning.

Just past Seaforth Channel, thousands of gulls were resting in the middle of Deer Pass. As I approached, they quietly and effortlessly lifted off the water, clearing a path for my boat.

Extreme tides mixed with the last of the winter's cold outflows marked the beginning of the herring spawning season. I have learned over the years that the Heiltsuk people look to the weather

The Heiltsuk village of Waglisla
(Bella Bella), facing east towards
the Coast Mountains.

first to predict when the herring will spawn. The gulls were here to feed on the tiny white eggs and the remains of the wounded fish left over from bubble-netting humpback whales. The forty-tonne whales, which had just moved inshore, dive below the herring and exhale in a full-circle formation, trapping the herring in the middle of the "net" formed by the resulting bubbles. White-sided dolphins, sea lions, and the many other predators of herring had also moved inshore to feed on the small, silvery fish.

I throttled down and peered into the clear water. Soon it would be milky white as the sperm of the herring mingled with the fertilized eggs, which lodge themselves on kelp and other seaweeds all along the coast.

I waited for the light of day; after seventeen years here, this remained my favourite time of day. The moon was still mostly full as it disappeared into the west, casting a purple glow over the snow-capped mountains surrounding Dean Channel. The day before, I had found tracks near the Fish Trap Pack den site. The wolves had returned to their denning island and would soon come down from the forest to feed on the eggs.

THE FISH TRAP PACK

About a month later, I watched as eight wolves glided by in single file on the trail that skirted the water's edge. It was a classic wolf trail, narrow and well worn.

Two of the wolves, including the alpha male, broke from the line and came to have a look at me. I was sitting in the same place I had sat many times before, under a large, old red cedar. I didn't move but just stared at the tree beside me, which was covered with dangling grey strips of bark interwoven with green Methuselah's beard, its lower boughs draped in brown eelgrass carried in on the tide. This tree had seen countless tides well before the first Europeans landed in North America.

Somewhere in the maze of branches above me, the ravens stopped calling; the other wolves must also have stopped, nearby. Head down and eyes looking up, the alpha male sniffed my boot tracks under the tree. I looked down so that he couldn't see my eyes. The second wolf watched him closely, waiting for his reaction.

The ravens egged the wolves on, as the Romans did the gladiators; the birds eagerly anticipated taking part in the

bounty of the next kill. Their presence added an edge to the moment, as I suspected was their intent.

The wolf patriarch checked me out most mornings. He was a gorgeous wolf, whom I called White Cheeks because of the distinctive white sides of his wide old face. He shared the main colours of his pack: ochre and sepia highlights on the ears and back, with black and silver bands running down his side.

It seemed that the purpose of his visit was to verify that I was alone and in my proper place. Three metres (ten feet) before my spot, he effortlessly jumped up the bank and quietly rejoined his family. The second wolf accepted his judgement of me and, joining the rest of the pack, continued up the trail to the den. Once more they had gifted me status, allowed me inside the perimeter.

One of the yearlings broke from his siblings and circled back, poking his nose through the dense salal bushes. It was Ernest. He was full-grown now but retained traces of youthful curiosity and the slight clumsiness of a yearling. He tiptoed so close that I could see the insides of his moist nostrils as they opened and closed, smelling everything about me. Ernest and I knew each other well. A barely discernible squeak from up the trail, almost a wheezing sound, followed by a low bark from one of the adults farther along in the trees, drew him quickly back to the pack. I had witnessed this ritual, this daily entrance to the den site, on most mornings over the previous few weeks, and it never failed to excite me.

Close to 450 kilograms (1,000 pounds) of combined weight, the wolves moved silently, like shadows of the rain forest, over barnacles, moss, and rocks. Not a scrape or a breaking twig was to be heard. They passed by so close to me that I thought they might touch me. As I glimpsed the blur of a paw on the moss or a huckleberry bush shaking slightly as a soft ear rubbed against it, I felt I was witnessing a great presence.

The mother, Urchin, was in the den, and I had not seen or heard from her for more than a week. She was graceful and strong and the alpha female, the mother of the last four litters of Fish Trap pups; this litter would be her fifth. When I first met her, more than seven years earlier, she was a yearling with strong brown and red highlights, but now she had

37 }

White Cheeks, leader of the
Fish Trap Pack and always
protective, checks to ensure
that I am in my proper place.

The Fish Trap Pack takes over the intertidal zone shortly after moving to their new rendezvous site from their higher-elevation birth-den location.

The main entrance to a wolf
den, which typically would
consist of one main den
connected by underground
tunnels to secondary ones.

turned largely silver and white from age. With experience, it is possible for observers to estimate the age of a wolf by the amount of grey or silver on the muzzle.

Her absence could only mean that the pups had been born. She was using the same den she had used for three years running, the same den in which many of these adult wolves had been born.

I wondered how many pups there were this time. She had given birth to an average of five pups each year. Remarkably, every one had survived until the fall season. Although more than one breeding female in a single pack has been recorded, this phenomenon is rare in unexploited wolf populations, and I have not observed it in this or any other pack.

From my vantage point, I could clearly see the first hundred metres up the river. Two of the wolves took up positions on either side of the water. They were brothers from the same litter, and I called them the Sentries. It was tough to tell them apart from a distance, but one had an unmistakable, scythe-shaped scar running across the side of his rust-coloured muzzle.

The two, now in the prime of their lives, had been serious even as pups, four years before. They took up their strategic

riverbank positions according to their morning ritual; nothing got in or out of the valley without their knowledge. Although tolerant enough of me, they seemed to be the least social of the pack.

The tide had turned and was starting to rise. It would soon cover this pack's namesake, an ancient stone fish weir built by Heiltsuk ancestors; the wolves made use of it to catch salmon in the fall.

Four sandhill cranes moved slowly and silently under the sweeping branches at the upper reaches of the tide line. When these normally gregarious birds with their rattling, prehistoric calls go silent, it is the time of year when the wolves give birth. The wolves themselves go silent at this time—not just this pack, but wolf packs all along the coast. They are also more cautious now; they will rarely respond to howling but will warily investigate intruders from the darkness of the forest. It is the one time of the year that they seem less the aggressive apex predators than cautious introverts. It is a survival strategy that has served coastal wolves and many other species well when they are caring for their young.

It was the stealth season, when reproduction and care for offspring demand all the attention of the family. Instead

Stealth Season

of spending their days howling, playing, and wandering, the wolves were focussed on providing security and food for the mother and the newborn pups. Every pack member was committed to this goal. If any one of them made a mistake, the den site could be discovered and an entire generation of the Fish Trap Pack put at risk.

Wolves choose den site locations in the core of their home territory, where there is less chance that wolves from other packs are in the area but also where many different and easily accessible food resources are available. They do not want to have to travel far from the site. The Fish Trap Pack den site was in a good spot, a small, steep-sided valley, close to a little lake. Nothing could enter or leave the valley without the pack's knowing well in advance. Deep burrows dug under the base of a massive red cedar kept the wolves dry.

The pups would emerge from the den three weeks after birth, and by fall they would be travelling with the pack. Until then, the family was at its most vulnerable, because it was anchored to one location. Flight was not an easy option, so the pack would do everything possible to avoid conflict. Knowing this made me even more grateful that the wolves had accepted me. Although I had been to the den site many times before, I wouldn't go there now and risk displacing them. The pups were just too young. The site was crucial to these wolves, and forcing them to move newborn pups would put them in danger.

Through my binoculars I could see Three Legs, or TL, appear briefly at the tree line. She glanced my way for a moment and then scanned the rest of the estuary. TL's rear left leg had been seriously broken just above the knee some years before. It appeared to have been a compound fracture, probably from an unplanned meeting with the hoof of a deer or maybe from a skirmish with a bear—maybe even from a bullet, but I didn't think so, because she was too trusting of me from the outset to carry that kind of history. The leg had atrophied and was now a useless appendage that hung listlessly below her haunch.

A Russian proverb says, "The wolf is fed by its feet." One would have thought, then, that TL would be at a significant disadvantage in providing for herself. But not among wolves, who, like humans and a handful of other highly social animals, can work co-operatively, capitalizing on division of labour.

A member of the Fish Trap Pack has a seal tail in its mouth. Wolf packs that have "haulout" rocks in their territories show higher levels of seal in their diet.

I had first seen TL here, in this same watershed, broken leg and all, five years earlier, an average wolf lifetime before. Although wolves living free from human hunters can live to be more than ten years of age in the wild, most die before they are five, according to L. David Mech in his book *The Way of the Wolf.*

Statistically, the odds were against TL's still being alive even with four working legs. But on this day she looked healthy and well fed, and though she was forced to traipse across some of the most onerous terrain on the planet, I suspected she was alive because of the food-rich environment and a relative lack of human persecution.

TL's speciality was the pups. When she first broke her leg, she became the designated stay-at-home pup-sitter while the rest of the pack was out hunting, and she had assumed this task ever since. This year she would be looking after the fourth litter of pups I had seen her raise.

As role model/caretaker, TL gave the pups the freedom they needed to learn about the world but also protected them from danger. With TL on sentry duty, a misguided wandering bear, dispersing wolf, or human with a fishing rod or gun would not have made it to the pups. I had to be on my best behaviour to win her trust over the years.

Now, incredibly, in this 150-square-kilometre (60-square-mile) home range territory of islands and forest-covered mountains, the pack arrived like clockwork in the same spot, within the same week of the same month, as they had for three consecutive years, to await the birth of the pups.

Although the Fish Traps stay close to the den site at this time of year, their full territory is situated in the core of Heiltsuk territory and is composed of numerous islands covered in red cedar, cypress, hemlock, and yew forests. In the rich, wet, flat areas surrounding the lakes and estuaries, the tall, silver-barked Sitka spruce forests dominate. Granite escarpments rise out of the rain forest, and at the base of these cliffs are caves where Heiltsuk ancestors still lie on wooden platforms. The wolves use these caves for protection in the winter.

Three main islands form the core territory, with ten smaller ones in the vicinity. I have frequently seen the wolves swimming from island to island, searching out deer and other prey. And although ten salmon-spawning rivers

Stealth Season

provide a significant portion of their fall diet, the territory is also rich with beaver, river otter, black bear, and waterfowl. The Fish Traps have the advantage of three main seal haulouts, or areas where seals rest on rocks.

The Heiltsuk refer to the core of the Fish Trap Pack's territory as "the Gateway." It is the entrance to many of the traditional wintering villages and ceremonial houses for the tribes and remains a rare part of the globe, where one can look over ancient, unbroken rain forest as far as the eye can see, a view that has not changed for the last five thousand years. It is almost a miracle that it still exists, given what is happening to old-growth forests on the rest of this coast and throughout the world.

The last fog of the day clung to the north-facing valley bottoms. It would burn off soon and give way to sunshine. An hour went by and nothing stirred; even the gulls, eagles, and ravens, which had fed all morning on herring, had now left the bay. The wolves had settled into the den site, where they would remain for the rest of the day. The sun felt good after the chill of the fog and warmed even my perch on the rock.

THE DRONE OF an outboard engine signalled that Raincoast wolf project researchers Chris Darimont and Chester Starr would be here shortly. I crossed the estuary as the boat pulled into the bay. Chris was already scanning the shore and saw all the tracks.

"Hot?" he asked, knowing the answer.

"Very," I responded. "Everyone was in attendance, and it looked like they made a kill to the south—at least, they came from that direction this morning, and they looked well fed."

Chester pulled out a smoke and smiled a toothless grin but said nothing.

I knew Chris was happy for me, but I also knew that he wished he had been here on such a special morning, especially this early in the season; the sedges were barely knee-high, and the days were still getting longer.

Chester and Chris split up the estuary, collection kits in hand, and went about their research. As I sorted out my film from the morning shoot, I was pleased that I had some empty canisters to show for it. Over the years, I have had countless early morning starts that led to not even one image. Photography was not the only reason I was out here, but even so, the lack of photographs could get a little frustrating.

I could see Chester and Chris at work up the river. Chester, also known as Lone Wolf, had just turned fifty. Chris was a fair bit younger than Chester's oldest son, meaning that the two men were more than a generation apart. Yet out here they were like brothers, and glaring divergences in background made no difference. I had never heard either of them question the validity of the other's perspective on wolves or the natural world.

In many ways they represented a fusion of Western science and tradition-based knowledge. Chris is one of the brightest, most dedicated scientists I know. His painstaking attention to detail made me nervous; his work was filled with lists of lists and measurements. In the lab or out here in the field, nothing went unrecorded in his Write-in-the-Rain notebook, and Chris validated all his work with scores of peer-reviewed science papers.

Lone Wolf's notebook, in contrast, rarely came out. And he was a man of few words. Chester's family is part Kitasoo from the village of Klemtu and part Heiltsuk from the village of Waglisla. When I first asked around Waglisla about who might be interested in working in the field on the wolf research project, every person I talked to directed me to the soft-spoken outdoorsman, tracker, and trained archaeologist who liked to travel alone. He was indeed the lone wolf.

He usually wrote down just the essentials. His tradition of knowledge transfer, after all, relied on oral transmission. Written records were a foreign construct, but he was patient with all of Chris's record-keeping requests. In fact, Chris confided that the notes Chester did take were the most detailed of any of the researchers' and often pointed out things that others missed.

Chris's yellow notebook would fill quickly with notes, data entries, and field observations. Chester worked in the moment, taking everything in and then mentioning it all in passing when he met Chris upriver—things such as deer bones from a recent wolf kill, maybe a sockeye salmon curiously mixed in with all the pinks, perhaps the flattened ground of an old village site, a culturally modified tree, an old cedar tree with a plank removed long ago for a big-house construction. Things that a casual observer could walk by countless times and not notice.

"It's kind of like my home," said Chester when I asked him what he thought of all the note taking. "I pretty much remember everything I see. And if I forget, I just go out in the boat and see it again."

I understood what he was saying. It would be strange for me to suddenly wake up and start documenting everything that was happening in my own home. Things that are so familiar rarely benefit from such documentation. We tend to record experiences that are foreign to us.

Traditional ecological knowledge and Western science are complementary. Chris has said that the two world views could even be synergistic, yielding more than expected when they merge, and that Chester's perspective has catalyzed his own thinking.

The parallel knowledge emerging from the two views is fascinating. For example, the Heiltsuk believe that two types of wolves exist within their traditional territory: the timber wolf of the mainland watersheds and the smaller coastal wolf, which is more closely associated with the islands. Although we did not wholly reject the idea, Chris and I were skeptical that two types of wolves could live so close together. And how could Chester differentiate between these two forms, especially from a boat in a channel almost a kilometre wide?

As it turned out, our skepticism was just a few millennia behind. Genetic and ecological research now shows that there are indeed two types of wolves here, mainland wolves and island wolves. Some of the differences between the two are fairly obvious, such as diet. The mainland wolves have access to mountain goat and moose, whereas the island wolves subsist on more easily accessible salmon, seal, and the occasional beached whale or sea lion. Other differences did not become apparent until the genetic samples made their way through the laboratory, and the results were startling.

The research team sampled extensively from six packs around the Bella Bella area, with fair representation of mainland and island wolves. What they found, regardless of distance between the two types, was that island packs are more related to other island packs than to mainland packs, and

Two Village Pack pups, almost five months old, wait for their siblings to join them before exploring the tidal zone.

vice versa. It makes sense; why would a young wolf born and used to island life and island diet disperse to the mainland? It just wouldn't be as likely as heading to a vacant area on another island. Over time these wolves have diverged genetically and most likely morphologically. As Lone Wolf pointed out before the first genetic sample was ever collected, the area has timber wolves and coastal wolves.

The history of the rain forest wolf is as speculative as its future. The most widely accepted theory is that wolves followed the northern expansion of deer in continental North America after the Wisconsin glaciers receded about ten thousand years ago, a supposition largely based on the assumption that deer were the primary prey of wolves and that the entire coast was covered in ice.

However, what if some of the outer islands provided ice-free refugia for wolves? Grizzly bear remains recovered from Prince of Wales Island in southeast Alaska date back 35,000 years and predate the last glaciation. Dr. Tom Reimchen at the University of Victoria and other scientists have identified distinct coastal and continental black bear lineages that might have been separated from each other

for 360,000 years, suggesting geographic isolation caused by ice-free refugia on the coast. The hoary marmot, ringed seal, Steller's sea lion, red fox and even a separate genetic lineage of sockeye salmon have been associated with such refugia; all of these are potential prey for wolves.

If research eventually provides evidence that wolves existed here throughout glaciation, it suggests a much, much older wolf association with the marine environment than is currently believed or understood.

AS THE FULL heat of the morning hit the estuary, Chester and Chris moved farther up the tree line, where they would record everything they saw. They would take samples of fresh wolf scat and place them in clear plastic vials, full of ethanol, for genetic analysis.

These tiny vials of genetic material would be sent around the world. Some would go to the University of California in Los Angeles, some to Uppsala University in Sweden, where researchers would painstakingly extract DNA from the cells that slough off along the intestinal tract. Besides distinguishing the mainland wolves from the

Stealth Season

island wolves, the DNA would identify individuals. Combining these data with information about where and when the samples were collected, the researchers would be able to describe where neighbouring packs' territories began and ended and how movements might be affected by mountain ranges or open ocean, availability of food, and season. Essentially everything that one could want to know about wolf movement could be traced according to what was contained in these small vials.

Much like Reimchen's work on bears, the DNA could also help reveal the evolutionary history of these rain forest wolves. Subtle changes, or mutations, occur over time in the genetic code of certain individuals, and such changes are forever marked as these individuals pass on their genes. Each version of the code, with its different mutation markers, is called a haplotype. More related populations will have a similar set of haplotypes compared with more distantly related populations. In other words, researchers compare the profile of the number and frequency of haplotypes among populations to estimate how they mix with one another. As genes are passed, offspring inherit the same

haplotypes from their mothers, allowing the researchers to develop a family tree of sorts.

This works because the transfer of genetic material provides a marker by which evolutionary histories can be traced. Specifically, if a population is isolated from others and does not interbreed, it will show a different profile of numbers and frequencies of haplotypes. If, however, wolves of one area shared similar kinds and relative frequencies of haplotypes with wolves from another area, it would be concluded that they also shared a similar evolutionary history.

That's not what is seen on the coast. Rain forest wolves share some haplotypes with their interior cousins, but they also harbour many haplotypes that wolves from the interior do not.

The unique haplotypes exist because coastal wolves are isolated from interior wolves and have not interbred with them, thus allowing separate genetic identities to emerge. Another reason the coastal wolves differ from those elsewhere is that the rain forest of British Columbia remains the only place in North America where humans have not played a significant role in shaping the wolves' recent evolutionary history. Everywhere else, humans have killed wolves at such high rates that they have wiped out much of the animals' former genetic diversity. And it is this genetic diversity that may be the saviour of wolves in the future, especially in light of new diseases and a changing climate.

Chester and Chris also put scat into a bag to be shipped to the University of Victoria for dietary analysis. There the scat would first be heated in an oven to kill potentially harmful parasites, and then biologist Johanna Gordon-Walker would laboriously sift through the dried scat—identifying every feather, hair, and bone. She had become an expert in describing the differences between seal, sea lion, river otter, beaver, mink, deer, mouse, and mountain goat hair.

Still another lab, at the University of Saskatchewan, would receive some of the scat. Veterinary pathologists there would search for diseases that the wolves might be carrying. A serious threat facing coastal wolves is that diseases may be transferred from domestic dogs at villages and logging sites. And, as crazy as it sounds, migratory birds that stop over on the estuaries may bring new diseases from, say, a pig farm in the southern United States. This could be devastating to the isolated wolves of the coast.

Scat was not the only thing Chris and Lone Wolf collected. Like dogs, wolves shed in the spring, so the research team had placed barbed-wire hair collectors on the wolf trails. In many ways, hair tells a better dietary story than a thousand scat samples. By studying the amount of marine-based chemical isotopes, Chris could differentiate between wolves that are marine specialists from those that focus on terrestrial prey. Fortunately, wolves shed their hair once a year, and a strand of hair is a dietary calendar of sorts.

Together, Chris and Chester were like a *Canis lupus* forensic cleanup team, tracing everything the Fish Trap Pack had been up to in the previous few weeks or longer. In addition to the traditional-meets-Western approach, their work was an amazing union of natural history and advanced laboratory research. The real beauty of it is that not a single wolf needed to be touched, harassed, or harmed. In fact, it was not necessary for any researcher or lab technician to even see a wolf.

By the end of 2006, the team had collected more than four thousand scat samples, walked more than five thousand kilometres (three thousand miles) of transects, collected more than twelve hundred wolf and prey hair samples, and located thirty-three den sites between Knight Inlet and the Alaskan panhandle. Chris had worn out a dozen pairs of gumboots and filled more than one hundred tattered notebooks.

I was envious of the time Chris and Chester spent together. The life for me, as a photographer, an observer, is a solitary one by necessity. Wolves don't particularly like to be around more than one person at a time. And more people mean more noise, more smells, more problems. Being alone with wolves is essential. It's like hitchhiking; you always have a better chance of a ride if you are alone.

The tide had covered the estuary, and I joined Chester and Chris for the trip up the Don Peninsula, where we had heard that the Village Pack was denning.

THE VILLAGE PACK

The scat was fresh—I could smell it as I walked along the wolf trail. Standing still and searching the ground, I finally located it, lying right up against the base of a tree. How had the wolf moved into a position to place it there?

Wolf scat is often found in strange locations. It might be on top of a stump or a rocky outcropping, on the bend in a trail, even at the top of a mountain—anywhere conspicuous enough to make the point and boost the signal. Wolves have an "anal sac" that is full of hormones and other chemical cues, and scat is the primary tool, among many, to mark territory, so it needs to stand out.

The moss around the base of the tree also carried the distinct yellow-green colour of many urine stains; this was clearly a scent post. Past the tree there was another fresh scat, though two fresh ones in a row are rare. This information coupled with the strong smell indicated that I was very close to the den site.

I could hear Gudrun Pflueger and Chester talking on the VHF radio, so I turned it down to a dull murmur. Gudrun, or Goody, as she is also called, was talking a mile a minute, as usual. She is always so full of excitement that she kind of gushes more than talks. Lone Wolf occasionally got in a few words of reply: "Yup, uh-huh, okay, Goody."

Goody had worked for three seasons on our wolf research. An Austrian-trained, world-class endurance athlete, she was as close as one can get to having the stamina of a wolf. For a noninvasive study that relied on footwork, she proved to be an irreplaceable researcher. She covered

Five Village Pack pups snack
on barnacles, crabs, and other
intertidal treats as they explore
the marine zone of their territory.

extreme distances without breaking a sweat, and she jogged along with a fairly large pack and chatted away with as little effort as if she were sitting in a coffee bar.

Because of her stamina, Goody was often assigned the most challenging of cross-island wolf transects, especially in the more difficult deer population surveys. These transects, randomly positioned throughout the landscape, were designed so that deer pellets could be counted along a straight compass course. Running through a tangled maze of devil's club, salmonberry, and many other thorny bushes, the researchers counted deer pellets and thus estimated the density of Sitka black-tailed deer—the main prey for wolves, especially during the winter months. Few people have been on top of as many peaks of the central Coast Mountains as Goody, and no one, I am assured, has done it so happily in search of deer and wolf shit.

We had covered a fair amount of ground since morning, and I knew from listening in on the progress of the rest of the field crew that they, too, were closing in on the den site. It's like being in the hub of a wheel; all of the trails eventually lead to the den site. Wolves are secretive, and den sites are the most important locations for wolves to keep secret. Sometimes I have criss-crossed a river system countless times, knowing that I was close, but never located the den. Sometimes the wolves are just damn hard to find, but when they are feeding hungry and growing pups, the trails get worn quickly.

I have found enough den sites to recognize the common characteristics—shelter from prevailing southeast winds, access to fresh water, distance from anything human, central location in home range territory—and I can usually locate a den after two or three days of work.

I crouched to make a closer inspection of the scat. It contained mostly meat and what looked like deer hair and large, sharp shards of bone. I picked at it with a branch, and the small hoof of a fawn appeared. As I gazed at the base of the tree, I suddenly realized that its trunk was an upright corner post and that the den site was among the remains of a big house, part of a very old and very extensive traditional Heiltsuk village site that stretched to the north adjacent to the beach. All along the oceanside as far as I could see were corner posts in various states of decay. The crossbeams,

Stealth Season

long since fallen, commingled into nurse logs surrounded by a latticework of roots. Huge Sitka spruces in straight lines grew out of the fallen structures.

The perfect lines of trees in a grid design were in stark contrast to the randomness of the rest of the forest. Although these spruces were huge trees, I knew they were probably not as old as they looked. The nutrient-rich midden of fish remains, clamshells, and bones, proof of thousands of years of human occupation, provided excellent nourishment that would make for elevated growth. The land was flat as far as I could see through the underbrush, and on this coast, really flat ground simply does not exist without help.

On the forest floor, I could see small sticks that had been chewed, small, dark holes at the bases of the trees, and small-diameter scat everywhere. The place was ripe with fresh wolf shit. The pups were here, and I knew that I had to leave. The wrens had stopped singing. It was too quiet.

As I made my way towards the ocean, two little grey pups emerged from underneath a confusion of fallen cross-beams. They didn't see me and went tumbling into another hole. At the same time, two more came running around the

corner. They dropped into the holes and popped out again like jack-in-the-boxes. The base of the log house was a network of holes and tunnels, and they were moving so fast that I couldn't tell whether I was watching a dozen pups or just the same two.

Finding a den site is the holy grail of wolf finds, but finding an ancient village site of such magnitude has to top the chart. There was a feeling here of very old human and wolf relations.

The human precontact population of this part of the Great Bear Rainforest region is estimated at about thirty thousand people. Given this high density, at least when compared with the present human population of about four thousand, it is not surprising that wolf and human sites would overlap. But the convergence of dens and ancient villages is beyond coincidental. More than 50 per cent of the den sites that we have identified are situated either within an abandoned village site like the Village Pack's choice of den location or in very close proximity.

Humans and wolves are top-level animals in the ecosystem, and they have similar requirements: shelter from

A wolf carries meat back to the den for the pups. A pack is only as strong as its last hunt, and a pack can easily consume a deer in a matter of hours.

winter storms, access to fresh water, and access to food sources such as salmon, deer, or mountain goat. This convergence between humans and wolves is probably what gives wolves such a prominent role in First Nations culture.

In the creation story of one of the founding Heiltsuk tribes, a wolf fathers the first children of this group. One child remains a wolf and serves as a protector of the people. His siblings stay in their human form and create many of the gifts to the people, including winter ceremonials, big houses, and salmon. The mother marks the wolf father with ochre paint, giving him a reddish tinge that is still common to the wolves of this area.

A lone wolf began to howl just in from the tree line, reminding me that it was time to leave. This season was starting out well. The team had already located three separate packs' active den sites within the core study area; this would be the fourth. Summer and fall looked promising.

THE NEXT DAY, Chris and I made an early start. Chris wanted to get a pup count, and I was keen to do some video work around the youngsters. We loaded his fibreglass

Stealth Season

speedboat with all the usual gear—canoe, tripods, spotting scopes, cameras, lenses, collection equipment, rain gear, anchors, radios, food, and survival gear, to name a few items. Farther back in the mountains, radio contact was sketchy, and it might be a few days before someone happened along to help out if we got stranded.

As we approached the den site, Chris slowed the engine. We would idle in the last one and a half kilometres, anchor the boat downwind, and canoe the rest of the way to an observation spot we had picked out the day before. When Chris asked for the binoculars, I handed them to him and followed his gaze. By squinting, I could pick out something in the water, but it was too far off to identify.

"Black bear," said Chris. I heard the anticipation in his voice. The bear was swimming directly for the den site, on the same beach where we had been the day before. We were pretty sure the bear did not know that the pack was just inside the tree line.

The bear was halfway to the beach. It took him only a few minutes to cross the channel, which was one and a half kilometres wide. As streamlined as a blimp, a bear can swim incredibly fast. This one's powerful legs were pumping away underneath the water, and we could hear him exhaling in loud, watery snorts. Sound travels remarkably well over water, and we knew the wolves could not miss huffing like that.

We watched the bear clamber onto the shore and, like a dog, shake the water off his coat. This was getting interesting. We knew from Chris's initial diet samples that the wolves had a significant amount of black bear in their diet—less than 5 per cent but still a considerable amount, taking into account that the average bear is not going to give itself up easily.

Something stopped the bear, and he cautiously moved back to the water's edge. His body tensed as he frantically sniffed the air.

Barry Lopez, in his superb book *Of Wolves and Men*, talks about the "conversation of death," in which the prey seems to willingly give itself up to the predator. He says that a form of communication takes place, perhaps a negotiation of some kind, in which the predator and prey each agree to play out their roles as ordained by nature. It is as if the prey sacrifices itself for the greater good. But this potential prey

A lone black wolf watches
me or, more likely, smells me,
half a kilometre away.

A black bear stands up on its hind legs for a better view of the estuary. While the relationship between coastal wolves and bears remains largely unstudied, one fact is that the black bear is on the wolves' menu.

Wolves in late spring and summer often look thin, but usually this is because they have shed their thick winter coats. One prerequisite for den location is access to fresh water.

was not a deer, moose, or goat, an ungulate with high biological reproduction success. This was a large, healthy bear, and I don't think it was overly interested in the greater good of this pack of wolves.

I had seen wolves take down deer at the water's edge enough times before to know that prey are often vulnerable in the intertidal zone. That crucial transition between land and the safety of deep water is a weak link for prey. It's like being chased on foot and suddenly seeing a bicycle: you know that if you can hop on that bike and start pedalling, you will be home free, but those crucial moments of getting on board and picking up enough momentum are when you are vulnerable.

The previous summer, my wife, Karen, was swimming with our son, Callum, in shallow water near an island across from our home on Denny Island when a deer came bounding out of the forest and ran straight into the ocean right near them. Seconds later a wolf followed. It bounced through the shallows, jumped onto the deer's back, and locked its jaws around the animal's neck. The struggle lasted only seconds, and within two minutes the wolf had dragged the small deer back into the forest. By the next day,

nothing remained except the rumen, or stomach, and blood on the rocks and moss. Callum was only three, but he still talks about that day.

Species that are on the wolves' menu, such as bear, deer, moose, otter, and beaver, can almost always outswim wolves in open water. But the time it takes the prey to get to the water and up to speed gives the wolves a last chance for a successful kill, and they often give everything they have to take it. The wolves might also use the water to soften the blow of flying hoofs. And by latching onto the neck or flank of its prey, a wolf only has to hold on and let the water assist in the killing. I had seen this pack make kills at the water's edge before.

The wolves lined up at the fringe of the trees, the pups safely stashed in the security of the den hole. Maybe the bear did not want to turn his back on the wolves; maybe he did not think he would make it back into deep water.

No longer stiff and cautious, he ran towards the wolves and disappeared into the forest. The wolves followed. Even from our distant location we could hear the crash of large branches breaking. The sound went on for a few more minutes. Then quiet returned.

Stealth Season

Wolf kills are often made close to the shoreline. This Sitka black-tailed deer was unsuccessful, but prey often can outswim wolves and try to escape to the relative safety of the water.

Fifteen minutes later the pack erupted in howls. The wolves did not sound anxious or aggressive but, rather, composed, as if everything was now okay. Perhaps the howls were a signal to the pups that the danger was over.

We decided to leave. If the wolves had felled the bear, we did not want to scare them off the kill. We would return the next day.

The following morning we canoed in and set up location about three hundred metres north of the kill site. After about forty minutes, two pups came out of the forest and started exploring and playing around the tide line. They were soon followed by the rest of the pups. There were six altogether, which is about one and a bit more than the average litter size. There was no sign of the adults.

One pup, a red-ochre female with great big ears, started following the water's edge, moving back and forth as the slow ripples lapped the smooth stones. She made her way along the shore until she was about four and a half metres from where we sat. Clearly she had followed this route a few times before and had never encountered anything like us. She moved a little closer but kept looking back at her siblings, unsure whether she should investigate further or not.

I took a few photos of her, and at the sound of the shutter she did a little jump and then trotted back to her siblings. It was a very nice welcome to the Village Pack.

The rising tide ushered the pups back into the tree line. While Chris moved farther north to collect more samples, I went in to investigate. We still did not know whether the bear had survived. I walked down the beach to where the bear had fled and ducked under the thick entanglement of huckleberries and into the dark of the forest.

I didn't need to go any farther than the thick understory of the forest edge. The place looked like a cross between a butcher shop and a barber shop. Thick clumps of black bear fur, with large chunks of flesh still clinging to some of it, were strewn all over the ground, and bones, hair, and broken branches were scattered everywhere. The bear had tried to make it up into the safety of a spruce tree that I had marvelled at the day before—the one growing out of the corner post at the village site. Judging by the scene before me, I figured the wolves had made the bear pay for entering their home site. I turned to leave as the pack once again started to howl. I didn't need any lessons myself.

I shouldn't have checked out the kill site so quickly, but

Wolves frequently use the shoreline at low tide as a travel corridor. There they also have a chance of scavenging a beached whale or sea lion.

the information was useful. Chris's dietary analysis tells us that most packs on the coast kill at least one or two bears a season, but it does not tell us the age of the bears. Did the wolves prey only on the old or sick or just on the young?

Here we had had direct observation of a large, healthy, adult black bear in its prime—maybe 160-plus kilograms (350 pounds)—killed almost at contact by the Village Pack. I gave Chris the news and knew he would be back with Chester to collect plenty of fresh genetic material once the wolves had finished their meal.

After seeing a big bear in its prime ripped to shreds, I wondered why wolves are so timid with relatively defenceless human beings. Some of the researchers, including Johanna and Goody, whose combined weight could barely break the 110-kilogram mark, are veterans of numerous close-range encounters with adult wolves at a den site. For some unknown reason, they can enter a site safely; the adult wolves consistently back off and temporarily abandon the pups.

Wolves can take down a full-grown black bear or a 450-kilogram moose. I have watched wolves surround and attack a healthy adult grizzly bear. Moose and bears alike are infinitely more capable of defending themselves and inflicting

serious damage on an attacking wolf than any unarmed human could ever hope to be.

I have walked into dozens of occupied den sites, sometimes by accident, sometimes to set up remote cameras. Every time, the same thing happens: The pups are left behind, and the adults move 50 to 150 metres into the forest and begin to howl. The adults are angry about the intruder and concerned in case the pups don't stay down in the den (they rarely do, since they are so curious), but they do not attack me. They do not do what they easily did to that black bear more than twice my size.

So why do wolves submit to us? Is it because they have learned that humans pose a different kind of threat, and therefore they fear us? Or is it because of something that used to exist between humans and wolves? Is it a social relationship that has been handed down from the time that humans and wolves lived together? I don't doubt that the ancestors of these wolves lived with the ancestors of the Heiltsuk people here. When these wolves let us into their lives, are they waiting for us to rediscover that relationship?

On *the* Edge

SUMMER

Companion's bow cut through the black water, leaving a trail of bioluminescence. It was an early summer night, and the northern lights shimmered in pulsating aqua-green waves above Mount Keyes. Towards the east, somewhere at the foot of the mountain, the Fish Traps were hunting in the dark.

The pack had just moved from their mountain-sheltered lakeside den and re-established their old rendezvous sites, a combination of nursery and playground, near the ocean. The pups were already well set in their daily routines of exploration and foraging. They supplemented the prey that the adults brought them with barnacles, mussels, clams, and any other critters they could find at low tide.

False lily of the valley and Indian paintbrush adorn the base of a Sitka spruce tree. The Raincoast Archipelago is a windswept, rugged, and spectacular wilderness.

I had followed them through the summer months for a number of years now and knew that their routines would serve them well until the salmon returned. And as much as I wished to stay to observe them, I was also keen to move offshore and spend some time with a completely different wolf society. I would be gone from this part of the coast for most of the short summer season to visit the Surf Pack. When I returned, these pups would probably have tasted salmon for the first time and would be twice the size they were now.

All sails were aloft, and I hoped that the wind would strengthen and carry me beyond Cape Mark before dawn. With luck, it would continue strengthening with a south bent to keep me going northward.

The forward port and starboard running lights cut the dark like small headlights. As I entered Queen Charlotte Sound, they seemed so small. I felt very small out here, too.

Once I was offshore, though, there would be fewer navigational dangers such as logs and debris. Running at night was a bit risky, but *Companion* was a multihull with no ballast, meaning that I couldn't sink even if all three hulls were damaged. The worst that could happen was the loss of my bottom gear—prop and shaft—from a submerged hazard (this has happened twice), but I would still have sails, and I had a backup outboard motor to use in an emergency.

Whatever the risks, I can't think of a sweeter sound than that of a boat under sail at night. Aboard *Companion* is one of my favourite places to be, and over the many years and the tens of thousands of kilometres we have travelled safely together, I have developed a strong affinity for this old craft.

Companion and I were tested out there. Storms had left few trees along the shoreline, and those that remained were gnarled krummholz. Islands like Calvert, Hunter, Aristazabal, Compania, Banks, and Dundas form the exposed backbone of British Columbia's north coast, but they are interspersed with many more island jewels. The wolves seem to prefer the northern extremes of these islands; the topography provides habitat for salmon systems, and the exposure to the prevailing southeast storms is not as extreme.

The swell began to rise and the wind freshened from the south as I made a few more turns on the mainsheet

On the Edge

winch, eased back into the cockpit, and slowly turned to the north. If all went well, I would be sailing for more than twenty hours, but the adrenalin associated with sailing off-shore, alone, would easily keep me awake throughout the voyage. The conditions were too good to waste on sleep, and I was eager to reach my destination.

A vast silhouette of countless islands—I call them the Raincoast Archipelago—disappeared into the eastern hori-zon. These islands are the first line of defence against the unrelenting winter storms that descend on the Great Bear Rainforest.

Although the Fish Trap Pack was not many kilometres to the east, at least as the crow flies, I was quite certain they did not know what surf was. The outer coast of northern B.C. is an extreme landscape, and the wolves that claim this uninhabited archipelago are very different from the inner coastal wolves.

Each island is an independent experiment in ecology and evolution. As scientists are learning, islands here that are separated by only a few kilometres of open ocean can host animals that differ radically from each other. How they make their living, what their movements are, even how they look—all can be different. And the geographic isolation that creates this biodiversity can increase its vulnerability.

From flightless birds to trusting mammals, the world's most notorious examples of extinction have occurred on islands. Here movement among land masses is less fre-quent. If something goes wrong—catastrophic disease, fire, effects of introduced predators—an island population (or species, in the case of large islands like Australia) can be wiped out. In isolation it can't be "rescued," meaning that additional individuals to replenish the population simply are not available.

Ian McTaggart-Cowan, one of the most distinguished early ecologists in North America, conducted some of the first field studies of these islands in the early 1940s, many of them focussed on the influence such isolation had on popu-lations. His study animal was the deer mouse, and he noted that mice on different islands also looked very different; they had taken another evolutionary course. His work was so pio-neering that it is still commonly cited today in scientific lit-erature, which is rare for work more than sixty years old.

Sorrow, a member of the Surf
Pack, was one of the friendliest
wolves I encountered and seemed
quite content to spend entire
days with me.

McTaggart-Cowan and his team documented a lot of the basic ecology of the coast. They showed the presence of many species, even conspicuous mammals, whose presence had previously been unknown on islands (to Western scientists, anyway).

McTaggart-Cowan was also a wolf man, and what he saw on the coast impressed him. Prior to his coastal research, he conducted the first work on wolves of the Rocky Mountains. Although wolf studies were tangential to his focus, he observed that coastal wolves were very different ecologically, morphologically, and behaviourally from the wolves of the Rockies.

The home of the Surf Pack is a diverse biological treasure chest, an explosive union of ecosystems that at first glance appears to be more of an unholy clash between terrestrial and marine environments. Waves surge in from the deep Pacific, incessantly pounding on rock, sand, and forest. Meanwhile, the trees expand their roots into the tidal zone, gripping and crushing rock. The two worlds converge, retreat, and are rarely at peace. This intertidal battle accounts for the superb biological richness.

And that was the very reason that I was there. Charlie Russell and Maureen Enns, two Canadian bear researchers, searched the globe for a population of brown bears that had had minimal human contact, and they ended up on the Russian frontier, the remote coast of the Kamchatka Peninsula. They spent eleven seasons living on the volcano-dominated peninsula with the huge brown bears in an effort to observe what a truly wild population was like—bears that had not been conditioned to guns and garbage, to people and their trappings. It was what I also searched for in wolves, and fortunately I did not have to travel to eastern Siberia to find them.

In the lee of one of the islands, after twenty-odd hours of sailing, I dropped anchor in eight metres (twenty-five feet) of gin-clear water. Shining a spotlight below, I could see sand lances and bull kelp swirling about in the current. I made sure that the anchor was well set; this spot would be home for as long as the southeast storms held offshore. I could hear the Steller's sea lions just off my port side, and the ambient roar of the surf filled the wheelhouse. The sounds and smells were familiar. I immediately fell into a

deep sleep, my clothes stiff from salt spray, tomorrow just hours away.

My eyes were still shut tight when the howls penetrated the boat. I had to think for a moment to remember where I was, but soon, above the sound of the surf, I could hear them. The wolves were still there.

I scrambled on deck and looked through the binoculars. From this distance, they were barely discernible from the sand, but I soon picked out six pups playing on the south side of the beach, running together at the foamy water's edge, the adults spread out across the length of this tidal river estuary, each one facing into the wind, their sides to me, and howling in turn. Six adults and six pups.

This was their ritual; every morning at daybreak, like clockwork, they would run together, as close-knit as any family could hope to be. This was their bonding time, and it was a joy to watch them out in the open and to see all members alive and well.

By the time I rode the canoe through the surf and walked up the valley, the wolves had long since left the beach for their two upriver rendezvous sites. The pups were now more than three months old and were travelling some one and a half kilometres (a mile or so) to the beach and back to their home sites every day.

The pups' current rendezvous sites were scattered no more than three hundred metres from the den site where they were born. One of the rendezvous areas was in a patch of scattered spruces, the other under the hollow of a large red cedar. The pups frequently moved from one to the other. It is a good strategy to have multiple rendezvous sites, or safe zones, for as their range increases, so does the opportunity to come into contact with potential predators. Den sites also accumulate disease agents, so moving to new ones may assist the pack in avoiding them.

Rounding the last bend in the tidal area, I saw the usual cast of outer coastal avian characters: a few sandhill cranes, some green teals, a Steller's jay, a kingfisher, sandpipers, and, of course, the ravens. They all moved about slowly, indicating that nothing exciting was happening with the wolves this morning.

Then I saw some movement down along the river's edge—a head popped up, snapping at a horsefly. Four adult

The adults of the Surf Pack face into the wind and all the smells that it brings on an early morning walkabout.

wolves were sleeping among the round boulders strewn about the flats. I had missed them because, each curled up head to tail, they looked just like the rocks.

As slowly and quietly as possible, I sat on the rocks with my camera and tripod. About a half hour later, the rising tide began to turn the wind around, and I knew my scent would now be heading their way.

The alpha male, in full summer coat (meaning not much of one), suddenly stood and stared directly into the shifting breeze. Although the coastal wolves, especially the outer coastal inhabitants, lose most of their coats and look pretty thin by late summer, this wolf, the leader of the Surf Pack, did not. Most wolves are built lean, like long-distance runners, but this wolf, whom I called Bob (for "big old boy"), had a chest bulging with muscle and attitude. He had a wide, scarred, dish-shaped face—he had been around the island, so to speak—and deserved respect.

He was also extremely protective of the pack and was the least sure of me and my intentions. The other three adults, taking his lead, also stood, now wary, and slowly stretched their legs while intermittently looking across the flats towards me.

On the Edge

On the Edge

I knew that I looked just like one of the boulders and that unless I moved I could not be seen. Wolves don't seem to identify stationary objects very well; they wait for movement. A human head moving slowly two hundred metres away can be picked up easily if the wolves are looking that way. I usually have to stare through binoculars and move only when a wolf is not gazing directly at me. Trying to move around while staring through binoculars has its own set of challenges, but it works.

I knew that the wolves would soon locate me by their sense of smell, however. Olfactory sense in a wolf is somewhere between a hundred and a million times more acute than ours, and there are few tricks that work to avoid being detected by smell. Wolves not only smell us from impossibly great distances but also smell what we had for breakfast—yesterday.

I settled in for what was about to happen. I had been in this situation before with this pack, and though I was pretty sure of the outcome, it still was unnerving because so much of my future stay here depended on its going well.

The wind was now consistently onshore, blowing harder and exaggerating my presence. The two remaining adults of the six I had seen on the beach came trotting out from the forest edge, about two hundred metres away. Equally attentive, they joined the other four, and the pack moved directly towards me. No question now, they had located me and were approaching to investigate.

Fanning out and then coming together about twenty metres away, they picked up speed and came straight at me. They gathered momentum, their trotting turned into a gallop, and then the gallop became a run. Within seconds, their tails went from relaxed to straight up, they held their heads high, and their ears were forward on full alert. I could see the guard hairs on their backs suddenly stand on end.

It was a hell of a welcome party, designed to flush me out and put me to chase. Not a chance. I would rather swim to China than run from a pack of wolves.

This was a crucial moment, and I knew that how I reacted in the next few minutes would set the stage for the time to come. If I screwed up, I might as well leave today. I had done it before, lost my nerve at the wrong moment by not trusting my instincts. If I stood up, wanting more space, they would be disturbed. I would be too dominant,

As pups play and explore on
their daily beach walk, one of them
contemplates making a playful
lunge at their constant travel
companions, the ravens. Although
many species of birds have
been found in wolf scat, raven is
notably absent.

and they could not tolerate that kind of presence in their home site, not with pups around. They would abandon this place and would not let me so close again.

The rush halted. Right now I was at the gate, on the perimeter, and if they accepted me, I would have access. If not, they would avoid me, and it would be very difficult to build trust again, especially with the alpha breeding pair.

The pack slowed down and was all around me. I was half crouched, half sitting on my rock as Bob came towards me, head held high above mine, then circled me tightly from only a few metres away, smelling everything, my tracks, my backpack, taking it all in. He walked a full circle, his feet seeming to barely touch the ground. He was tense, and suddenly his whole body, as strong as that of any wolf I have ever seen, sidestepped away, as if surprised, almost as if a new scent had triggered some distant bad memory.

The other wolves, who had been watching their leader circling, jumped to attention. Things were not going the way I had hoped and expected. Had something bad happened to this pack while I was gone? I had seen no human tracks on the way in and didn't think anyone else had been

here, but it was impossible to tell for sure. Wolves are shot from boats; the animals travel long distances and can run into many problems. But out here it would be exceptionally rare for them to have contact with people.

The ravens had moved down to where the excitement was and were flipping back and forth above us, calling out. Their cries added to the tension. As the pack became more aggressive, Bob started to bark with a strong, guttural, mono-syllabic woof; it was as if he were convulsing, regurgitating food, but all that came out was this loud bark from deep in his chest. I lowered myself a few centimetres and spoke to him softly: "Hey there." My voice was a familiar sound to the pack; I immediately sensed that their muscles relaxed slightly. The hairs lowered perceptibly on Bob's back, and his tail dropped slightly. I think he remembered me.

I raised my head and spoke again, slightly louder this time. On cue, the rest of the pack relaxed. The older female—not the mother of this year's pups but maybe the breeding female from years past—abruptly sat down on the salicornia and began to scratch her lower stomach with a rear leg. She snapped at a horsefly and then rolled onto her

On the Edge

back, her legs raised to the sky. The other, lighter-coloured female sat back on her haunches and stared, ears forward, alert and pointed right at me, her aggressive disposition melting away to curiosity. This wolf would become the friendliest of the pack, and we would spend much time together in the coming days. I named her Sorrow for her sad-looking eyes.

Bob was now, thankfully, satisfied with the situation and trotted back upriver, where the rest fell in line, single file. Their business was taken care of. They stopped about fifty metres away, back in the boulder field, and promptly scattered about, lying down on the cool, damp ground with their tails touching their noses. This is the litmus test of a wolf pack's trust—when they fall asleep in your presence.

I had been gone a while, but they remembered me. I would be allowed back in.

THE FIRST TIME I encountered the Surf Pack, I assumed that they were on the move and that I had only intercepted them as they travelled from one feeding site to the other. This turned out to be far from true. On closer inspection,

it became clear that these were not ephemeral trails used occasionally, or even seasonally, but were old and well worn. I had stumbled upon a well-established home site, one that had been used by this pack's ancestors for many generations. I walked through the coastal fringe spruce forests on trails that were centimetres deep in the dirt and moss or worn right down to white rock on weathered granite slabs. It takes a lot of constant foot action, especially when the foot is a fairly soft paw, to make permanent trails like that.

Over time, I learned that there were two main travel routes for the Surf Pack—the shoreline perimeter trails, which went from beach to beach to beach, and the trail that bisected the island, connecting its two main salmon rivers and accessing the forested slopes of the interior of the island. Many trails branched off them, but these were the main arteries.

The home range of the Surf Pack is very different from that of the Fish Trap Pack; it is almost as if they are from different continents. In contrast to the Fish Traps' home territory of many islands covered in lush rain forest, here the terrain is constantly changing and exceptionally diverse;

The last rays of the sun highlight the tree-draping *Usnea* lichen, or old man's beard.

the Surf Pack's domain includes wide open bog forests, green intertidal fields of delicious salicornia, sand dunes, and moss-covered spruce dripping with green lichens. But the big difference is in diet.

The Surf Pack scats I found dotted across the beach were full of smaller marine-associated life and thus closely resembled river otter scat. The scats also included small feathers from seabirds and shorebirds and large feathers from cranes, geese, and herons, as well as small fish bones—maybe from a cod, a quillback, or a copper rockfish that might have rolled in on the surf—and large vertebrae bones from salmon, a small tuft that looked like rodent hair, a fragment of an eggshell, mussel and clamshells, a shard of bone that felt heavier than deer bone, maybe from a sea lion or seal. These wolves make their living in a very different way from the mainland or inner coastal wolves.

On a clear day I could look across the strait and see the Coast Mountains of the mainland. Those same wolves that I had seen hunt mountain goats alongside wolverines only fifty kilometres away preyed on creatures that the wolves out here did not even know existed. Nowhere else on this planet do two types of wolves, practically in sight of each other, have such different diets.

Moose and mountain goat are just two of the species that are not found out here. These wolves are true saltwater wolves, unique outer coastal predators sustained by the sea that carried them here. Did they come here because of these prey opportunities and empty real estate, or were two wolves blown off course as they travelled to one of the more familiar islands and then pushed here by the strong offshore currents?

ONE DAY AS I sat watching the wolves, an otter popped out of the river just downstream from the pack. The otter was clearly on a mission, and it looked to me, based on the direction and the speed it was travelling, to be a suicidal one. It was around noon, and the pack was still sleeping, scattered throughout the estuary. From my vantage point, I could see only the tips of their ears. From the river otter's vantage point, just a wall of grass could be seen as it sort of side-loped along, as otters tend to do on land, but this one danced right smack into the middle of a dozen wolves.

On the Edge

Sorrow heard the splashing and got up, stretching her front legs out before her, and immediately located the otter. She must have communicated something to the others, because everyone woke and looked at the otter, which by now was pinballing in its crazy little sidestepping fashion from wolf to wolf.

At every turn it made, the otter found itself heading towards another wolf. The pack was sitting up and watching with interest. I could sense that although the otter never fully stopped, it did suddenly understand its situation—it had run all by its little self straight into hell's kitchen. It picked up speed, loping now with purpose towards a river tributary only nine metres from Bob's careful gaze.

And the wolves just sat watching it, not moving, as it entered the tributary and escaped into the water. Not a single wolf even bothered to get up on all fours. I was amazed. The otter had bounced through the pack, an absolutely perfect and easy afternoon snack for whoever wanted to take the trouble, and it had not elicited a single wolf's serious attention. In a glance, the pups realized that the adults were not up for a chase and promptly went back to chewing sticks and digging holes.

To me, letting this brazen otter go unchecked seemed inappropriate behaviour to be modelling in front of the pups. The whole scene was just not wolflike.

Paul Paquet later told me a number of similar stories about wolves and prey from his studies in the Rocky Mountains. Once, when he was studying a den site occupied by a pack of wolves, a deer walked directly in front of the den only three metres away from the nearest wolf; they merely watched it.

Barry Lopez, in *Of Wolves and Men*, eloquently describes what can be the sacredness of hunting from a human perspective, and he questions what the hunt means to wolves: "Can hunting be regarded as a sacred occupation among wolves? Is there a mythic contract acknowledged when wolf meets prey? We can only ask the questions and guess. Communal hunting probably *is* the social activity that makes wolves hold together in packs. *Sacred* is not the right word, but hunting may have overtones for wolves that we cannot appreciate."

Lopez continues: "Here are hunting wolves doing many inexplicable things (to the human eye). They start to chase an animal and then turn and walk away. They glance at

A pup tries to finish off a
fence post put up some
one hundred years ago by
Scandinavian homesteaders.

a set of moose tracks only a minute old, sniff and go on, ignoring them."

I understand what Lopez is struggling with. It seems that the hunt is a ritual that takes mental preparation. It is not a random activity but one that takes place at the appropriate and designated time.

WITH THE SURF PACK

The pups of the Surf Pack spend a good part of their day playing below the high-tide line. It is through play that they sort out their position in the hierarchy and determine who might one day become the alpha male or female.

But some play seems to be just that. Usually the Surf Pack's six young siblings made their way down to the beach ahead of the adults. Often I watched from the deck of the *Companion* with high-powered binoculars. The games generally started when something washed up on the beach, the same sort of something that on any other day probably would not be given a second glance.

One day the chosen treasure to fight over was a tangled piece of weather-hardened bull kelp about six metres long.

The alpha pup, who was larger than his siblings and consistently dominant over them, had the booty in his mouth and was running down the beach with the thick end firmly held in his jaw, the thinner end trailing along after it like a slithery snake winding its way around the rocks. As he headed for the safety of the tree line, but just before the last sight of the kelp disappeared into the forest, the siblings caught up and in one big pile grabbed hold of the prize and the tug-of-war started. I was surprised at how strong that kelp was, but then I remembered that west coast Natives wove sun-dried kelp strands into lines for catching halibut, which can weigh many hundreds of kilograms.

The siblings were pulling the kelp slowly back onto the beach. This tussle went on for hours while the adults, long since finished their cruise of the open beach, watched from their perches on top of the large boulders at the river mouth.

One evening, I watched the pups of the Surf Pack justify my name for them. They surfed. They started by running in circles on the beach; just beyond them, the surf rolled in with crests as high as two and a half metres. With perfect timing, the youngsters climbed the sand dunes

On the Edge

formed where the river meets the incoming tide and then bounded over the edge, jumping on top of each other and rolling down the sand face to end up submerged in saltwater and foam. Moments later they scratched their way back up the hill, shaking off the sand and white foam. Covered from head to toe in the bubbles, they looked like little manicured French poodles, pompoms and all.

A few days later, halfway up the tidal estuary, I found myself again holding my breath with Bob. The wolves were scattered about in various locations, sleeping the day away. I didn't want to change my position, as everyone else seemed quite comfortable at the moment, but I was losing my ground to the incoming tide—already my boots were covered, and the water was rising fast. Having lost more than a few pieces of equipment to the ocean over the years, I do not like operating camera gear over salt water.

As I worked my way towards the trees to a new spot under the shade of a rock, Bob's head came up. Maybe he didn't like the fact that I was more hidden now, or maybe he just felt compelled to do something. Without standing, he arched his large, powerful, beautiful head back and yawned a great big yawn that easily could have encompassed a watermelon. Such yawning is classic displacement behaviour, signalling not tiredness but anxiety. I could see the black bands that highlighted his lower mandible. He looked directly at me, and then a low, slow howl came from deep in his belly. About three-quarters of the way into it, just when I thought I should leave him in peace and get out of there, his head dropped. He was asleep before it hit his paws.

None of the other pack members even raised an ear in interest; this was a lethargic, sleepy afternoon. I leaned back in the moss, closed my eyes, and joined the pack in sleep.

This family did not always act as one cohesive pup-rearing team. Two wolves spent less time with the pups than the others; these two were a male and female, both possibly beta or second ranked in the pack hierarchy, but not yearlings or omega, the lowest ranked. They looked similar, so most likely they were siblings. Unlike the other adults, they seemed happiest to spend time with each other or alone and often went many days or longer without visiting the pups.

At first I thought the two were just happy to be away from the boundless energy of six little wolf pups, and

Sorrow's curiosity and her playful
and gentle disposition made her a
welcome companion.

maybe I was partly correct. But soon I realized that they played another role as well. By positioning themselves away from the pack and in strategic locations, either up on a bluff with a clear view of the flats below or at the very outside of the oxbow bend that anyone or anything entering the valley would have to pass, they served as sentries. Their behaviour was very similar to the Fish Traps' defence system.

I established a relationship with these two guardians from the beginning. I had to, or I could not get even within sight of the pack's main rendezvous site. I had a careful and deliberate routine. I always took things slowly, stuck to the same drill, crossed the river at the same point, sat in the same spot day after day. At first they weren't sure that I could be trusted. I could see it in their facial expressions and in their eyes. It was their job, after all, to alert the pack to intruders, but I was so confidently projecting that this was normal behaviour for a human that they eventually accepted the routine.

The older female I called Sorrow was especially friendly. She was sylphlike, graceful, and petite, and she had silver fur. Although her large, deep-set, auburn eyes looked sad to me, she had the nicest disposition one could hope for in a wolf.

Most days, once I was settled into my observation post, she would wander over to me and lie in the deep grass close by, sometimes only six or nine metres away. She kept an ear out if I was doing something that to her was strange, but basically I became just another resident in the estuary, like the cranes and geese or the kingfishers. One time I awoke to find her sniffing my boots—she must have been watching me for some time before I woke.

I learned that if I really wanted to know what was going on around me, I just needed to study the direction of her ears. It was like having a surrogate pair of eyes or a nose that actually worked. A wolf's ability to smell is nearly impossible for us to comprehend. To put it in perspective, imagine being blindfolded and travelling through the rain forest trying to locate deer purely by smell. This is unimaginable for a human but is the daily routine for these wolves. I looked where Sorrow's ears and nose were pointing, and I got the jump on whatever was coming our way.

The wolves had an uncanny ability to communicate what appeared to me more like intuition than anything else. Time and again I thought that I was close enough to the pack to see and hear every move and sound they made, yet

On the Edge

somehow, without my hearing anything or observing any movement on their part that would indicate a call to gather and move, one wolf stood up and headed off down to the beach or into the forest, and one by one, slowly but surely, the rest followed. It was as if they all just by coincidence came up with the same idea.

One day, when I was watching the pups playing on an island in the middle of the river, Bob started howling. The rest of the pack joined in, and the pups added their howls whenever and however they could. The cries cycled through all possible howl types, from plaintive to whining to commanding to fire-truck-siren loud. Bob ran in circles, head down, scruff on the back of his neck standing. He kept a very high-pitched whining going, as of a squeaky wheel going round and round. The pups knew that meant something was up.

Unbeknownst to me and the pups, out of our sight the adults had made a successful kill near the den site. The pack howls started about a quarter of a kilometre to the east. Usually the pups and the ravens ignored the chorus, but this was different. One by one, the ravens slowly headed upriver.

The howling continued just as the six pups reached the tree line. Suddenly the smallest one, a loner pup, stopped and ran all the way back to the den site. The howling continued, and it soon became obvious that the wolves were howling for the missing pup, which for whatever reason was not leaving the den to join in the kill. The alpha mom came back out to the estuary, calling for the wayward daughter, but the pup did not respond, at least not that I could tell. Mom looked peeved and maybe concerned as well. She went back into the forest and to the kill site.

Twenty minutes went by and she reappeared, crossed the river and the length of the estuary, and again gave the very audible, wheezing, concerned squeaky-wheel call, the call that means business. This time the reluctant pup came out of the den site, ran up to her mother, and began to jump at her muzzle. She was still jumping at her mother's mouth when suddenly large chunks of red meat came flowing out, regurgitated by Mom and onto the ground.

Perhaps this little pup's intransigent behaviour had purpose. Perhaps the adults do not pay close attention to which of the youngsters get enough to eat and which don't. This pup had probably learned the hard way that because of her

Goose feathers lie on a classic
wolf trail: straight, narrow,
and purposeful. Canada geese
are prey for wolves.

size she often got the leftovers of the kill. With five hungry siblings to compete with, that likely meant not much food for her. This way she was more assured of a decent piece of the menu.

After studying the prey remains in both adult and pup scats at the same time, over two years at two den sites, the research team has learned that pups are raised on a special formula. Parents and older siblings save younger deer for provisioning of pups and feed on older deer themselves. One theory is that the providers are giving the pups the safest morsels; deer accumulate more parasites with age, and fawns therefore are safest for the pups while their immune system develops.

I did not want to disturb the pack, so I waited until the next day to inspect the kill site. It turned out that they had killed a beaver.

After a few more long days on the estuary without much more than an occasional encounter with Sorrow, I decided to stretch my legs and do some exploring. I knew that the wolves were spending more and more time farther upriver under cover of the rain forest. I left my camera gear behind, took enough food for the day, and set off.

I chose one of a series of wolf trails heading off to the east and followed it until it intersected with the other trails and became one very well-worn path. Starting alongside the tree line, it was narrow and old, a typical no-nonsense wolf trail. Within ten minutes, I found myself in a very large network of beaver dams. It was amazing that an active beaver colony could exist so close to the wolf den. I would have expected the wolves to take them out years ago, but the beavers had cleverly designed a defence mechanism that was working well. Nevertheless, these must have been some stressed-out, vigilant beavers.

The trail moved through the forest of mainly small-diameter yellow and red cedar with a large component of shore pine. The huckleberries were fat and plentiful, and I noticed a fair amount of ungulate browse on them. As usual, the cedar forests were doing best on well-drained slopes or ridges where bogs could not form.

I dropped down into a bog, located in the flat or slightly sloping catchments. The moss was so thick and wet that I could push a stick three metres down into it and still not reach bottom. But as long as I stayed on the wolf trail I was okay—this was an all-season route, and the wolves knew where the solid ground was. In wolf country where the snow is deep, wolves walk in single file so that only one has to break track. Out here on the wild edge of the Pacific Ocean, wolves do the same in the bogs to provide similar advantage. The difference is that these trails do not need to be rebuilt constantly.

Farther along I could see cedar stumps, an old fence post, and through the salal the remains of an old homestead. The main cabin, which was half fallen over, was made of hand-hewn cedar logs, squared off with dovetail corner joints, a nice piece of construction.

It was almost as if I were being taken on a historic route. Scandinavians tried to homestead here more than a hundred years ago, mainly because of rumours of gold and because the salmon canneries were close by. So many of these remote outer coastal places, where flat ground gave the seriously misguided impression of agricultural potential, were the scenes of failed attempts at homesteading.

At first it seemed like paradise for the new arrivals from Europe—rivers teeming with salmon, estuaries on which to farm livestock, a land of plenty to be tamed and exploited after dikes had been built to keep the ocean swells out of

On the Edge

the estuaries. The attitude was that everything was possible with the appropriate amount of sweat and perseverance. And, of course, a dose of European farming techniques.

All that is left of these dreams and hardships is what sat in front of me now. Rusted plows and shells of rotten cabins, and, if you look closely enough, you can see the rise of land where the dikes were built in an effort to hold back the tide. What were they thinking? Trying to stem the tide on a coast that annually receives more than three and a half metres of rain, surges backed by hurricane-force winds with six-metre tidal ranges.

The trail brought more surprises. As I dropped down into a spruce- and hemlock-fringed flat on the eastern shore of the island, I saw a number of day beds of the Surf Pack. Plenty of feathers of great blue herons were spread around as well, which was puzzling; a great blue heron would not be an easy catch for a wolf, and the nutritional benefits not worth the expenditure of energy. Then I looked up and saw that I was underneath an extensive heronry. The wolves would probably get an occasional snack from the nests when a chick fell out.

The trail headed off north and south. I chose the northern route, which meandered from white sand to grey granite to a path beneath the spruce trees. Mostly the beach walks were for high-tide traverses from headland to headland. After a hundred metres, the path made a decided turn towards a clump of trees perched on an islet, accessible to everything below at midtide. It was a well-used river otter haulout, one of nine the wolf trail would have taken me to by the end of the day.

Yet more wolf food lay ahead. The next cove over, the trail headed out to a point and to a small series of rocks that hosted about twenty-five seals, looking like bananas with heads and tails curved up to the sky. I suspected that the wolves probably swam out to the rocks at night and attacked under cover of darkness, when the tide had dropped and the seals were farther from the safety of the water.

Farther down, the trail dipped back into the forest to a large spruce housing an old nest of a bald eagle pair. Underneath the huge tangle of branches, sticks, rope, and plastic, I noticed a crow's nest. The crows feed on whatever delicacy falls through or over the edge of the eagle nest. Farther

This smallest of the Surf Pack
pups often travelled around the
rendezvous sites alone.

on, the trail veered to a couple of mink holes and to a cove choked with driftwood that would catch everything floating in when the outflow and northerly winds picked up. Perhaps a seal shot by a fisherman, a whale that died of old age, a driftnet full of fish—all of these things and more would end up in this cove.

This wolf walkabout was so consistently worn that the wolves must monitor the coastline continually. I suspect that in part the strategy is based on the simple premise that the more kilometres they put in, the better the odds of intercepting food; in part it also is based on plain, old-fashioned ritual and habit. And when wolves are on the hunt, they just need to move—it's in their DNA, their very being.

The final section of this trail took me from the salmon river spawning grounds on the east side of the island directly across the higher treed mountain slope, where I found many deer beds and pellets. The path eventually dropped down to the other salmon system on the west side of the island. This was a route with a dizzying array of food opportunities.

When I returned to *Companion* that night, exhausted, I measured the distance I had covered. It had taken me about ten hours to make the oceanside circuit and the cross-island trek. It was probably no more than about ten kilometres of distance. For a wolf familiar with the trail, it would take a couple of hours at most. Two hours, and within them more than twenty different food possibilities that I could identify.

ONE MORNING A few days later, I had a feeling that something new had happened. The ravens were louder, and they were low on the ground near the water's edge. A wolf wandered through the birds, scattering them into flying cartwheels. The ravens moved right back in again. The pack had made a kill on the shore near the beach.

After a while, Bob walked onto the tidal rocks with something large and fleshy in his mouth; then off he went into the woods. Ten minutes later the whole pack erupted into howls.

One of the pups left the crying family, went to the shoreline, pulled at the kill, and headed down towards me. When I saw what he had in his mouth, I couldn't believe it. Dangling out of the side of his muzzle was a fleshy arm about sixty centimetres (two feet) long and covered in big, white, round suction cups. It looked like an octopus arm.

I examine a giant Humboldt squid, a rare visitor to the coast of British Columbia that proved to be a surprising food source for the Surf Pack.

OPPOSITE: This scat is evidence that the Surf Pack consumes a lot of salal berries and huckleberries.

I waited until the incoming tide covered the area where the wolves had been feeding. After they headed upriver for the afternoon, I took the canoe in through the gentle swells and waves. When I reached the site, I quickly realized that the buffet was not one octopus but many giant Humboldt squid. I had never seen squid this size before. They must have come in on the high tide and become trapped in the old fish weir. The wolves had already consumed a number of them, and tentacles, big black beaks, ink, and squid flesh were strewn about the beach. It was a calamari feast.

Add a new marine species to coastal wolf diet. I suspect at least one of the diners got a mouthful of ink, apparently a Japanese delicacy, but I wondered what the wolf thought of it. I took a chunk of an uneaten squid for my own dinner that night.

ALTHOUGH I WAS there to observe wolves, I spent more of my time observing ravens. They are always with or near the pack, and they are easier to spot in the trees, wheeling through the canopy over the sleeping wolves.

By watching the ravens, I could locate where the wolves were napping. Often, when they were in full sleep mode, the pack spread out on both sides of the river. Some of them slept in the open estuary, hidden by the tall grass, and some slept under the forest canopy. It could be a chore trying to locate them all, but by watching the ravens do the rounds I could usually find them.

Wolves and ravens clearly have a reciprocal and mutually beneficial relationship. If the ravens were not around, it usually meant that the wolves had made a kill away from the home site. Besides snatching up any scraps left behind at the kill sites, the ravens also fed right alongside the wolves when meat was aplenty.

This pack's home site was very clean, largely because the ravens eat the scat. They especially like to clean things up around the den sites after the wolves have been eating marine mammals—this scat tends to be very smelly, and that seems to be a selling feature for the ravens.

Ravens are often the last ones to clean up a carcass after the wolves have made a kill. When they finish, they follow the wolf trails, eating the scat as they work to catch up to the pack.

Sorrow (left) and Bob turn simultaneously after hearing the foreign sound of my camera shutter.

How do the wolves benefit from the presence of the ravens high overhead? Just as I use the ravens to follow the wolves, the wolves probably use them to keep track of me and of any other foreigners in the area. The ravens also can cover much more ground than the wolves, especially helpful on the coast in the search for carcasses that might wash up on the beach. The ravens alert the wolves to the carcass, and the wolves break the tough outer skin for the ravens. Seals and sea lions have very thick skin and centimetres of blubber that would be very difficult for a raven to open up on its own.

Where wolves and ravens coexisted elsewhere—and the two were historically well distributed throughout the Northern Hemisphere—their relationship has been described in evolutionary time scales. In his seminal work *The Wolf: Ecology and Conservation of an Endangered Species,* L. David Mech writes: "Both species are extremely social, so they must possess the psychological mechanisms necessary for forming social attachments. Perhaps in some way individuals of each species have included members of the other in their group and have formed bonds with them."

I have watched wolves and ravens play together for hours, especially with the pups. Ravens delight in dive-bombing the unsuspecting youngsters, grabbing at their ears in mid-flight. The pups snap at them and stalk them when they are on the ground. Although they appear to try, I have never seen one succeed in catching a raven; I suspect that if they were ever successful, they would feel lousy about it. Surely it would be too much like killing a family member.

Chris Darimont's dietary study analysed more than 3,300 scats collected from the outer coast, from islands like this one and up into the Coast Mountains, and the results show that a wide variety of birds, such as sandhill cranes, Canada geese, ducks, and herons, are prey for wolves. But, surprisingly, not a single trace of raven has so far been found in wolf scat. Mech's analysis of thousands of scats in Alaska from previous studies revealed the same.

The pairing of wolves and ravens exists in non-Native mythology, also. In *The Wolf Almanac,* Robert H. Busch says: "... the god Odin kept two huge wolves at his side that accompanied him into battle along with two ravens, which tore at the corpses of the dead. The name Wolfram,

On the Edge

meaning 'wolf-raven,' became a great warrior's name, and to see a wolf and a raven together on the way to battle was thought to ensure victory."

Members of the Tlingit Nation, occupying the northern extent of the coastal rain forest, have been observing wolf-raven relations for so many thousands of years that it defines their own social organization. The Tlingit are divided mostly into two groups: Ravens and Wolves.

Intermarriage outside the clan traditionally was allowed with either Ravens or Wolves. But Ravens and Wolves were debarred from intermarriage within their own ranks. According to Tom McFeat, writing in *Indians of the North Pacific Coast*, "a Raven man must marry a Wolf woman, a Wolf man a Raven woman." The children of such pairs belong to the mother's phratry.

Although the raven is higher ranking than the wolf in the beliefs of the Tlingit, the services that these two animals provide each other is mirrored in Tlingit ceremonial structure. Within Tlingit society, only members of the Wolf clan conduct funeral ceremonies of the Raven, and at feast time only Wolves can serve and distribute property to Ravens.

I have learned a great deal about the First Peoples on this coast by watching wolves, just as I have learned a great deal about wolves by studying traditional culture. Considering the close historical relationship between wolves and people here, I sometimes think the wolves are waiting for the abandoned villages to again be occupied and shared.

ONE OF THE distinguishing differences between the wolves and me is that I already know what I will have for dinner tonight. Having lived on the boat for quite a few years and provisioned for long, open-ocean voyages, I always have many months' extra stores. Aside from the odd cache, wolves don't. They are only as strong as their last hunt.

Each morning the pack walked downriver, crossed six stone fish weirs, and gathered on a rocky granite point surrounded by white sand dunes and seaweed from the previous tide. Then they ran together on the beach, and I didn't usually go upriver until they had performed this morning ritual.

On one such morning, I could see some of the adults sleeping on the smooth rocks, feeling the first sun of the day. Suddenly a few of them got up and looked across the bay. The alpha male was trotting along the tree line towards the pack. He reached them a few minutes later, and they headed back the way he had come, all in a straight line. This was a rare sight.

Twelve wolves in single file only paces apart moved directly across the sand flats, heading for the coast trail that travelled up a small seasonal side creek off to the west. I had never seen the pups taken this far with such urgency, at least not in the middle of the day. Maybe the pack was moving location.

I had to decide what to do. To follow closely might spook them, and if they got up into the rocky escarpments it would be difficult to track them. Then I noticed two of the pups digging for something under a log. They lagged a good way behind the pack, and it would be easy enough to follow these two without being detected. Soon the pups took off in a hurry to catch up to the pack, and off I went, too.

I followed the tracks easily enough through the forest, but occasionally I could see the two little rascals up ahead meandering along, exploring everything that they possibly could. The trail took me into a grove of giant cedars—very short giant cedars, since nothing grows tall in this part of world, exposed as it is to such extreme weather. Many of the trees had had planks taken out of them by First Nations long ago, and cedar-bark strips could be seen everywhere. Judging by how the trees had grown around the edges of the plank strips, they were very old. Down through the skunk cabbage and huckleberry groves, the trail eventually spilled into a vast, open carpet of sphagnum moss and the occasional gnarled shore pine. I once cut down one of these tiny pines in my backyard while clearing a trail, and I had to get a magnifying glass to count the rings. I counted more than one hundred years in a tree only fifteen centimetres in diameter.

Five kilometres and three hours later, I was still following the pups. I was surprised that at five months of age they were travelling this far largely on their own.

On top of one of the larger hills that the trail skirted, I could look out to the north and see the grey-black sky deepening. The barometer had dropped quickly that morning, and I should have been at the boat and ready to ride the southeast storm north.

Ravens and wolves have forged
a mutually beneficial relationship
over millennia.

On the Edge

The relationship between humans and wolves continues to be celebrated as Heiltsuk youths perform the Wolf Dance in K'vai big house.

Rain inhibits wolves' ability to hunt, because scent will not travel as far, so the Surf Pack adults often stayed close to the pups during inclement weather.

It would be tough to leave not knowing where the pack was and why the pups were travelling so far. The easy life of summer, of pups being just pups, was almost over for the six little ones. Fall storms were becoming the norm, and the first cold outflows rolling out of the interior, bringing a deep chill to this island, were not far away.

The wind pummelled my back and my hood blew off. My camera lens and tarp flew into the air. Gone for good. Cold water splashed down my back, and as I looked to the sea I shivered at the blackness of the sky and the whitecaps racing across the grey water. I decided to turn back and make for the boat.

Pulling the anchor, I went aft and raised the sail; with a loud snap, it filled immediately, bringing *Companion's* bow around fast, her nose following the wind. Under a single storm sail, I surfed down the large swells, burying the needle at thirteen knots. White water and steep swells surrounded me. I heard things crash onto the floor down below, plates and whatever, but I didn't bother to look. I had to keep my hands on the wheel.

I wondered where the pups were. Had they rejoined the pack? Did they make it to the intersection of the shore trail and the cross-island trail? Were they on top of a kill, tasting meat, fresh blood, anticipating the salmon they were about to meet for the first time? I hoped so.

As I glanced back through the almost-horizontal rain, the Surf Pack's home melded into one dark band on the horizon. I hated to leave, especially now that my senses were becoming fine-tuned to this place and the daily routines. I was actually using my nose to inform my decisions, and by listening to ravens I could tell what the wolves were doing. But it was time to catch up to the Fish Traps.

Three hours after the kill
and the diligent attentions
of three wolves, only the
rumen (stomach) and
some hide remain from
an adult deer.

Salmon Forests

FALL

THE RAIN pounded down on the roof of my boat; it sounded like a gorilla banging away up there. At the same time, the storm-force gusts hit the boat broadside and almost threw me out of my bunk. But I knew that the salmon were now finally entering the forest and would get to their spawning grounds. And the wolves would be waiting for them.

This had been the driest season on the central coast since records had been kept—the driest for a hundred years. It was the first time that the elders in Waglisla had ever seen so many snowless mountaintops from the village. Even the lake-fed systems had not contained enough water for the salmon to migrate upriver, so they had been pooling at the mouth. Marine predators such as seals, sea lions, halibut, whales, eagles, and humans had taken their fill.

Wolves travel across water
as humans cross streets:
watchfully, but regularly. Wolves
have been found on islands
ten kilometres from B.C.'s
mainland coast.

I had been gone only a few months from the Fish Trap Pack's territory, but as I sailed up the inlet, I could sense that things had changed. It was strangely quiet.

WITH THE FISH TRAPS AGAIN

I stopped off to check the pack's den site. I knew they were absent before I had even set the anchor, and not just because I couldn't hear or see any birds in the forest. This den island does not have the salmon resources that the neighbouring one does, so before the salmon migrate upriver, the Fish Traps change location and swim across the channel to the spawning grounds. Islands that provide greater spawning density tend to support more resident competitors such as bears. The wolves thus avoid such islands in the spring and early summer while the pups are vulnerable.

The swim would be the longest the pups had taken so far, but I knew they would have made this aquatic journey with ease.

I was fortunate enough to watch the wolves make the swim one August. I can't usually observe every pack member at once, but they were all on hand for this move. They made the crossing in single file, with adults bringing up the stern and the bow of the wolf flotilla. Wolves swim across channels and inlets as we cross streets, looking both ways to make sure there is no oncoming traffic.

Many times I have come around the bend in a motorboat and unintentionally forced a wolf to turn around and abandon a crossing. Wolves need to be cautious, since moose and deer swimming between islands have been swallowed by orcas, and some humans armed with guns or gaffs would show little mercy.

When the wolves reached the opposite shore that August, they shook off the salt water and looked back the way they had come, waiting for Three Legs to arrive. I could only imagine how this new experience would be for the pups—a different set of islands, an expansion of their territorial world, a new playground to explore.

Back in the present, at the tide line I located the main trail to the den site. Just one fresh set of adult tracks crossed the mud flat; there were no pup tracks anywhere. Since thousands of tracks are typically associated with active rendezvous and den sites, it was clear that the pack had moved on in search of salmon.

Salmon Forests

Salmon Forests

I followed the trail along the edge of a small lake and approached the den. The forest understory on these islands varies from place to place, but I have found that wolves choose den locations in fairly thick vegetation, especially amid salal bushes, which crackle loudly, like rice paper, when you walk through them. It is almost impossible to approach the den site undetected.

The grass in front of the den was worn down to bare earth from the pups' play-fighting, sleeping, digging, and exploring. A maze of trails led in and out of the base of two large, gnarled cedars, which were leaning over the tannin-darkened lake. The tops of fallen cedars poked out of the water, their bases secured deep in the muddy bottom. Gulls made their nests in the secure, mossy tops of these partially submerged logs. A loon called from the other side of the lake.

I have been to more than thirty den sites on the coast, and although they are almost always located in beautiful spots, this one was especially so. It was exquisite. Although I am sure that the pack chose this location because of its access to prey and fresh water and for security and ease of defence, I still couldn't help thinking that a more aestheti-

cally appealing place could not be found. It appears that the needs of wolves, when choosing a den site, converge with our view of beauty in the natural world. But it also makes sense that we associate beauty with necessity and utility.

Trails radiated from the den. Large, round balls of green and red mosses, a metre (three feet) in diameter, were scattered about and served as perfect day beds for the adults. The roots of the multiheaded red cedar that provided the support structure for the den below were covered in teeth marks from the pups' insatiable desire to chew everything in sight. Among these day beds, rounded out from the warm bodies of the wolves, a rivulet of water cascaded down through moss and grass.

The den smelled dry and comforting. I shone a flashlight inside and felt the floor, which was hard packed, its edges covered with soft, shredded cedar bark, twigs, grass, and dry moss. Pup fur was scattered everywhere, and the larger guard hairs of the mother were snagged here and there on the roof of the den.

This site had been used as shelter for enough generations of the Fish Trap Pack's offspring to buff the root-formed entrance smooth. At a now-famous den site in the

Two of Ernest's siblings in the
Fish Trap Pack watch me from a
favourite observation post.

Arctic, radio dating of bones from prey suggested that it had been used by wolves over a seven-hundred-year period. I looked around at some of these cedar trees and knew that seven hundred years was not an uncommon age for them.

Deer bones and feathers from what looked like a Canada goose were scattered about outside the den. I wondered whether the wolves would use the den the following year. They seem to reuse den sites but only intermittently, since the pack changes its breeding female and each has her own idea of the ideal den site.

In southeast Alaska, Dr. Dave Person of the Alaska Department of Fish and Game located twenty-four den sites between 1993 and 2004. All were in old-growth forests and about a hundred metres from fresh water, and all were carved into the basement of an old tree. Only one of the dens was found under a fallen log.

Re-anchoring the boat across the channel, I launched the canoe and paddled slowly into the bay. The excited calls of ravens, gulls, and eagles could be heard upriver. Salmon were jumping all around me, boiling over on the surface like river upwellings.

After I reached the beach, I hoped that I could make it to my spot in the forest quietly, without causing a disturbance. Turning towards the forest, I stopped abruptly. Just past the golden and red crabapple trees, Ernest was sitting, waiting. As usual, he had placed himself right in my path, doing what Ernest does best—staring intently. There was no doubt that he recognized me.

After a quiet greeting, I circled off to one side of him. I knew I was not going to win a staring contest with the self-proclaimed gatekeeper of the pack. He had figured out at an early age (all of a few months before) that howling, growling, or generally being a wolf was not the way to discourage me and that alerting the adults did not elicit much interest.

I was pretty sure that the rest of the pack was pleased to have Ernest off their hands for a while, and if the odd-ball creature hanging out under the tree (me) was happy to entertain him, so be it. He was convinced that if the pack were to be protected, he would have to do it himself, so his general strategy was to boldly plant himself in clear view, sit on his haunches, and with both ears pointed forward and a furrow in his brow, stare directly at me without flinching.

I heard somewhere that a human can outstare any mammal on the planet, and I have found this to be true—except with Ernest.

He was a peculiar wolf even as a pup, frequently straying from the pack and exploring the estuary on his own while the other pups preferred to stay in a group. He did not seem to be positioned low on the social scale; in fact, when all the pups were together he appeared to be the dominant or alpha sibling. I think he just had an independent streak in his personality and liked to sort out things for himself.

Watching him from inside the tree line, I wondered what role he would carve out in the complex and very social hierarchy in this family. Were strength and physical dominance more important than intelligence and social aptitude? Ernest seemed to have it all. I suspect that, as with our best leaders, a combination of these traits works best. Maybe some pups are destined to lead from an early age.

The responsibility appeared enormous. The alpha provides leadership for up to fifteen individuals, maintaining pack cohesion, strategizing hunts, and deciding, usually on an empty stomach, whether to move in search of prey or to remain patient, hoping for a successful hunt. Leadership among wolves is complex, and it varies according to season and pack. What seems to be common, however, is sex-based roles. Top males lead other males, usually in association with hunting and defending territories. But many times, as I have found when close to pups, it is the ranking female who takes the lead in defending the pack. One of the most important decisions to be made is the location of the birth den, the core site for the pack for many months, and it is chosen by the breeding female.

Ernest, for all his quirkiness, was also downright beautiful, with dark, striped highlights in his coat, bright grey-white eyes, and a swagger that separated him from his siblings. I had never encountered a young wolf like him before. He had such an unnerving sense of earnestness that he made me feel a bit sheepish for intruding on the pack. But once I was past him and settled in my spot hidden away under the trees, I very much admired him for it. It was as if he were trying not to be one of the pack, even if just for a little while. This behavioural variation is what makes wolf observation so interesting; each pack has such a diverse cast of characters.

Wolves, unlike bears, do not have
the advantage of large retractable
claws so must lock their jaws
around the backs of the salmon.
Nonetheless, they are extremely
efficient salmon predators.

Ernest's staring strategy was quite effective. We eventually agreed that I would stick to my one spot underneath a patch of spruces and hemlocks, instead of wandering around the fishing rendezvous site, and in return he would stop staring at me.

Although I wanted to push the boundaries of this relationship with Ernest, I know that a human-habituated wolf, one that naively, unconditionally trusts people, will not live to be an old wolf, not even in this remote part of the world. Wolves here arguably suffer the least amount of human persecution in all of North America—possibly in the whole world—but they still must be cautious.

Here, wolves are almost always killed indiscriminately and opportunistically. There are no limits to the number a commercial guide-outfitter can kill. Resident hunters are allowed to kill three wolves per year but do not need a special licence, though one is required for every other large-mammal hunt and even for geese and ducks. Because reporting is not mandatory in British Columbia, no one knows for sure how many wolves are killed by hunters, but it is estimated that about 2 to 3 per cent of rain forest wolves

throughout the Great Bear Rainforest die this way each year. Closer to communities and roads, the number killed rises significantly.

The temperate climate means rain forest wolves have a less commercially desirable coat than wolves in the colder Far North or the interior of Canada, so trapping coastal wolves is not considered economical. But if deer or bear hunters, fishers, loggers, or landowners see a wolf, they may shoot to kill. Some guide-outfitters advertise wolf hunts and pride themselves on their ability to locate and kill wolves. Many hunters I talk to say they are teaching the wolves a lesson—as if a dead animal, no matter how intelligent, can learn a lesson.

Some hunters offer the excuse that they are helping to boost deer populations by killing wolves; they call this "ungulate enhancement." But modern work on disease in ungulates has shown that populations without wolves and other predators are much more likely to suffer catastrophic disease outbreaks. In addition, wolves are obligate carnivores and must eat meat continually. If deer decline, so will the wolves, often quickly, thus allowing deer to bounce

A young Fish Trap Pack
member grabs a headless
pink salmon killed by one of
the adults farther upriver.

back in large numbers. This fact is an elementary truism of predator-prey dynamics.

On some of B.C.'s outer islands, smaller than the Surf Pack's islands, wolves can reduce deer numbers while maintaining their own population by then switching to a marine-based diet. Such a diet has its limits, however, and the wolves will often move to other islands or die.

When humans kill wolves, the wolf population will often retreat and reproduce in greater numbers. Humans form the only species on the planet that eats itself out of house and home. Easter Island and the Pleistocene overkill are two such lessons, but our current unsustainable existence on Earth may be the most relevant example.

AT THE RIVER

Leaving Ernest on the beach, I made my way to the south side of the bay, a couple of hundred metres from the river mouth, and entered the forest. A thin trail took me through devil's club and salmonberry bushes to one of the main riffles, the first shallow section that the salmon had to travel across and one of the wolves' favourite fishing spots.

The salmon pooled in the lower reach, churning and splashing, still strong. They were moving up from one rapid to the next, but the tide was dropping; the salmon that did not make it in time would have to wait in the pools for the next tide. They were tense as they felt their bodies getting heavier, pushing harder against the gravel with every centimetre of dropping tide. This was the first time in four years that they were not fully encased and suspended by water. The salmon were mostly pinks and chums, which can spawn in brackish water. But coho salmon can't, and they were already in the deeper pools upriver.

I looked around and saw that headless salmon were strewn about the forest floor, the bodies bearing the distinct puncture marks of wolves. Typically, the wolves eat just the heads. One hypothesis is that they are avoiding a parasite that concentrates in the guts of salmon. It's a parasite that also uses the bear and weasel family as final hosts (but does them little harm) and yet for some reason is deadly to canids in large concentrations; government agencies in the United States have even considered using the parasite to kill so-called problem coyotes. The wolves have evolved

Wolf-killed chum salmon await the next scavenger in the food web. The brain is preferred by wolves, possibly to avoid parasites in the stomach and for the higher nutritional benefits associated with this part of the fish.

to eat salmon by avoiding the viscera in which the cysts of this parasite concentrate.

As the sun dropped out of sight, the ravens became more active, flying between the salmon carcasses on the riverbank and a spot above a small rise across the water. A resident kingfisher patrolled the river's edge, hoping to scare coho fry from behind a log with its loud, staccato call. A dipper rhythmically bounced up and down on a smooth stone before diving into the current in search of a wayward salmon egg.

Then I heard something from up on the ridge, and a moment later the black-banded brothers appeared, the pair I called the Sentries, their heads held high as they looked towards me. I was mostly concealed, but I knew that they were aware I was in the valley; even if they could not smell me, I was sure that Ernest had given me up. Nevertheless, they both stepped out in full view and surveyed the river. They had almost identical markings—a dark belt of black across their chests; white patches on their cheeks and muzzles, surrounded by black penumbra etched with silver highlights; ochre and beige across the sides of their bodies. They were both stunningly beautiful, their colours uncan-

nily taken from the surrounding rain forest palette. Neither of them was the alpha leader, but I had my bets that one of them, one day, would lead the Fish Trap Pack.

The two wolves commanded this river as I had never seen a bear or another creature do. They both stared intently, as if with one set of eyes, at my position across the river. Some struggling salmon diverted their attention. Perhaps they were telling me that they knew where I was but also that I would be tolerated once again. I felt very fortunate. With all such encounters, I believe that a fragment of the trust that once existed between wolves and the First Peoples of this coast is rekindled, that I am witnessing the potential for humans to find their place again in the natural world.

I pressed the button on my camera, and the sound of the shutter brought both the wolves' heads up to stare at me again. It was unbelievable that they could hear that sound over such a great distance and with all the loud, running water between us.

The rainwater from the night before had brought many more salmon into the river, and fishing conditions were perfect. Soon the rest of the pack filtered out of the trees behind the brothers.

Salmon Forests

A wolf plays a significant role in transporting a salmon carcass from the river into the forest. Studies have shown that up to 80 per cent of marine-based nitrogen fertilizing the riparian forests comes from the remains of spawning salmon.

Salmon Forests

These two brothers I called the
Sentries were vigilant lookouts for
the Fish Trap Pack.

Salmon Forests

Salmon are considered the "foundation species" on the coast. Wolves are just one member of the salmon-consuming guild, which includes more than two hundred species, but as I looked down on the river, they stood out above all.

The pack's alpha male, White Cheeks, padded gingerly across the barnacle-encrusted rocks towards the river. If necessary, a wolf will sprint across barnacles or other sharp ground without a moment's hesitation, but a cut paw is not worth risking for salmon.

Sensing an approach, a pink made a dash for it across the riffle but was trapped between a couple of smooth, black rocks. Its battered tail whipped back and forth in a struggle to find deeper water. White Cheeks trotted towards the salmon and lunged at it, his tail swinging in the opposite direction to keep balanced, and with a single motion he locked his jaws onto the back of the salmon's head. With a quick twist of his neck, he carried the thrashing fish in his mouth up onto the riverbank. There he placed his front left paw on the side of the salmon's head to steady its wriggling front end and pressed his right front paw on the salmon's body. The salmon was a female, and a few round, orange eggs sprayed across his eyes and face.

White Cheeks thrust his mouth to engulf all of the salmon's head and lined up his carnassial or shearing teeth for the *coup de grâce*. Like a can opener, he began to cut out the brain. His own head almost grotesquely twisted, the salmon's head reaching the very back of his mouth, he stared blindly at the sky, and within seconds the entire brain cavity was cleanly severed, just where the gill arch began. The crunching of salmon bones was audible. White Cheeks took only a few chews, followed by a big swallow, and the salmon head was gone. The whole operation took about twenty seconds.

I have seen these precise and efficient actions time and time again in many wolves; this is not incidental or spontaneous feeding behaviour. Such skill, coupled with a consistently high success rate (wolves get their targeted salmon in at least 30 per cent of their attempts, which is high for any predator), gives us tremendous insight into the history of wolf-salmon interaction. This behaviour is clearly ancient; salmon are not a new resource for wolves.

The predator-prey relationship is as long as the time these two have spent together, and they have been engaged in a co-evolutionary drama. The size, shape, and behaviour

Pups eat barnacles during low tide. Marine-based and easily accessible food sources such as this may be one reason that pup survivorship is very high in the rain forest.

of these salmon have most likely been forged at least partially by the ancestors of these fishing wolves. In turn the wolves have most likely developed characteristics suited to preying on salmon—for example, their digestive physiology and coloration (the red-ochre colour seen in many coastal wolves reflects the colour of the seaweed adorning the rocky shores). And we know that this co-evolution is old, because of the high efficiencies and conserved behaviour in both species. These traits witnessed in predator-prey relationships take a long time to evolve.

The pack had now completely taken over the lower stretch of the river. Normally this kind of habitat would be dominated by grizzly bears, but they had been forced upstream to the less desirable fishing areas in the river. Only in the mainland watersheds do grizzly bears fully dominate certain parts of the rivers; on the islands where there are fewer grizzlies, they are more accommodating. I have watched grizzlies and wolves feeding quite peacefully alongside each other, and I have also watched them fight to near death over a single salmon. I wish I understood the relationship between the two species better, but I can only

surmise that by merit of their intelligence and individual personalities no useful generalization can be made.

One morning, just to the north of here, I watched a family of black wolves and a family of grizzly bears share the best feeding spot on the river. One of the bears chased a wolf that had just caught a salmon, which was thrashing in the wolf's mouth. The wolf dropped the fish, and the bear grabbed it. Suddenly three of the wolves charged the bear and the bear dropped the fish, which was repossessed by the wolves. This battle went on while the poor salmon kept thrashing around on the gravel and eventually died unclaimed by either side. When I walked through the forest later that evening, I found the wolves sleeping in the bears' fresh day beds.

Here at the ebbing river, by the next morning more than two hundred salmon had been caught. The remains of their carcasses left behind by the wolves would be quickly consumed by birds, beetles, and many creatures in between and would contribute important nutrients to the forest floor. The leaves and needles of streamside vegetation would benefit from these nutrients and would then nourish invertebrates,

Salmon Forests

which in turn would feed the next generation of salmon in the river. And, in an endless, elegant cycle, those salmon would feed future generations of wolves.

Pup survival rates in the rain forest are among the highest for wolves anywhere, most likely because salmon is a predictable and easily accessible food supply. The wolves do not choose salmon because other prey is not available; they specifically select salmon.

Chris's "resource selection" analysis shows how wolves select salmon over deer. In other words, the greater the abundance of salmon in a pack's territory, the more the wolves consume. It doesn't matter how many deer are around; wolves eat more salmon if their territory offers more salmon. The lives of wolves do not hinge just on ungulate prey, as is so closely held in established wolf science orthodoxy.

The islands provide a significant portion of the overall salmon-producing creeks on the coast, and although they are smaller in size and volume than the mainland glacier-fed systems, they also provide a disproportionate benefit for the species known to consume salmon. For wolves the small systems are especially important: bears have the advantage of large, retractable claws to catch salmon, but wolves have

to use their jaws and so are at a disadvantage when fishing in deep water. I have seen wolves dive completely underwater to catch salmon, but it is tough, and the success rate is lower than when fishing in shallower water. So although some of the larger river systems may support more salmon, they do not provide the same access for terrestrial wildlife because the water is just too deep.

Downstream at the fish weir, rows of rocks were piled on top of each other centuries ago in a long, undulating curve from the mouth of the river and across the estuary. The trap ended at the tree line, about two hundred metres from the river mouth. The three separate walls remain, although mostly torn apart long ago by the Canadian federal Department of Fisheries and Oceans (DFO), which at the time said that the Aboriginal people were destroying the salmon fishery by catching the fish at the mouth of the river; the DFO made the use of stone fish traps illegal. Today, only about one hundred years after declaring the weirs unusable, the DFO is experimenting with "terminal fisheries"—otherwise known as catching fish at the mouths of rivers.

Terminal fisheries make sense. If a salmon is caught in front of its natal stream, there is no mystery about where it

Opportunistic ravens stay close by as the Fish Trap Pack goes about an evening of fishing that can land more than two hundred salmon in a single session.

came from or where it is heading. If enough salmon have entered the river to spawn, then a responsible decision can be made about how many can be harvested.

In contrast, just a few kilometres from this river, the DFO allows unsustainable levels of fish to be caught in large, concentrated net fisheries. Before they can reach wolves, bears, and hundreds of other users, fish are taken by humans at rates far exceeding that of other predators. Dr. Tom Reimchen at the University of Victoria has calculated that non-human predators, on average, claim about 10 per cent of the biomass of their prey each year. In contrast, he has calculated that commercial fisheries commonly take 50 to 90 per cent. We have emerged as the planet's super-predator.

Although fishing fleets assume that they are targeting one large stock at the entrance to an inlet or farther out towards open ocean, the countless smaller stocks intercepted incidentally are being wiped out in the process. It is these small systems that fuel the greatest diversity of life on the edge of the Great Bear Rainforest.

An empty creek at the height of the spawn is heartbreaking to witness. It is like entering a river in the dead of winter.

Salmon Forests

The incongruous silence of a river that should be teeming with life is deafening. Because of the ecologically irresponsible commercial fishery, overall returns and genetic diversity (fuelled by the many smaller creeks) have been seriously compromised on B.C.'s central and north coast.

But for wolves, the fish traps make sense. Close to a hundred years later, the weir still works to hold back the upriver migration enough to give wolves a fishing advantage.

TL WAITED PATIENTLY for a salmon that was slowly moving upriver. The fish was looking for protection from all the other wolves as it swam right between TL's front legs. With one deft move, she latched onto its back. The four-and-a-half-kilogram (ten-pound) fish thrashed around, but TL limped up to the riverbank and ripped into it.

Based on all the studies of wolves available today, these wolves should prey on deer or beaver or some other large mammal, but this pack remains fixated on the salmon. Some days they catch more than fifty before 9 AM.

This level of fish consumption in wolves has never been documented before. Scattered accounts appear in books twice as old as I am, and an early paper from Alaska that used stable isotope analysis suggested that marine resources are important to wolves, though the authors could not be certain salmon was the sole marine input.

Chris, working with Dr. Reimchen's lab, tells a more detailed tale. Their modern isotopic work analysing the marine signatures in wolf hair has shown that both salmon and marine mammals provide tremendous marine biomass to wolves, and the amount of that biomass depends critically on where the wolves live. On the mainland, wolves make only about 25 per cent of their living on marine foods. Packs like the Fish Traps, on inner islands, split their diet fifty-fifty between deer and marine foods (and especially salmon). The Surf Pack and other outer island wolves are as marine as a wolf can get; up to 75 per cent of their nutrients come from the sea. Not bad for a "terrestrial" predator.

For the wolves, such a diet is good economics and sound

decision making. Why travel kilometres for a meal when a salmon swims right to you and your pups? Why risk losing a leg (or worse) in taking down a deer when you don't have to? Why pass up a predictable and nutritious food source? Salmon brains deliver high doses of omega-3 fatty acids.

The next day, Chris and Lone Wolf would arrive and count the salmon carcasses left behind. Wolf canine teeth leave puncture marks that can easily be distinguished from those of other salmon predators. Chris and Lone Wolf would slash each salmon tail with a serrated knife so that it would not be counted twice. They would note the number of fish consumed and what part of each was selected.

They would also check each salmon to see whether it had successfully spawned. They find that most salmon that wolves catch—as with bears—have at least partially spawned. Killing partially spawned males allows others to successfully compete to fertilize the eggs deposited by the females, leading, in theory, to an increase in genetic diversity in offspring. Chalk up another ecosystem service to the wolves.

Many people mistakenly attribute the leftover remains of salmon carcasses to bears. But bears lack the finesse and precision of wolves and do not sever just the head. Bears tend to make a mess of things, ripping the skin off, tearing the fish to shreds, and leaving a few scattered remains such as the jawbone and tail. The end of the salmon run, after all the fish have spawned, is a dynamic seasonal metamorphosis, almost a season between seasons. Post-hyperphagic bears disappear into the high country to sleep off the long winter. Only salmon bones remain on the ground, but in the water and soils the fish continue to penetrate the salmon forests at molecular levels. Millions of small eggs are safely covered and protected deep in the gravel beds below the cold water. The birds that feasted on the dying salmon disperse, until the spring herring season. All of this concentrated life centred on salmon now shifts survival strategies for the coming winter.

The shift is especially pronounced for the salmon-fishing wolves up and down the coast. The sun makes a very shallow arc in the southern sky; the season is drawing to its end. The days are short, dark, and cool. Any morning now, the pack and others like it will move farther into the interior of the island, targeting the slopes for deer. And for the first time since the pups were born, the wolf family will travel as one unit, exploring its winter territory.

Nomads *of the* Rain Forest

WINTER

WINTER WAS approaching, and storm-force southeasters were becoming so frequent that I had stopped listening to the weather reports. As one front after another swept down on the coast, it was simply best to be prepared for the worst at all times. The wind coupled with deep water and poor holding ground made for precarious anchoring, but at least the coast had been cleared of the fair-weather folks—the flotillas of yachts and sport fishers. Except for Lone Wolf, the researchers had gone south as well to analyse data and process samples in their labs. Even for the First Nations, it was a quiet time; the salmon were gone, and the herring wouldn't arrive until the early spring. I didn't expect to see anyone out here.

The snow line dropped closer to sea level every day, and the rains had cleansed the rivers of remaining carcasses. The bears had left the valley floor and were now above the snow line, inspecting their winter dens and readying themselves for the long sleep ahead of them. The wolves rule winter here, and some of these bears and their cubs would be attacked and killed in their dens before spring arrived.

As I walked through the forest, frozen huckleberries shattered like glass chandeliers. Deciduous trees stood like skeletons, their fallen leaves commingled with the decaying salmon now floating out in the bay. The geese were digging through the ice, eating last season's sedge roots. I thought a flock of crows might indicate a kill, but they were just busy dropping clams on the rocks to break them open. I could see a few marbled murrelets still coupled and now in winter plumage, and the last of the sandhill cranes had long since gone south. Only the year-round residents of the coast remained.

I had come to the watershed where the Fish Traps had spent the past three months fishing, and I found only silence. And no fresh tracks.

In the previous few years a late run of coho had brought the wolves back, and I hoped that had happened again. But after walking the entire spawning length of the river and not finding a single fresh salmon, I knew that the run was finally over. I would have to wait for the snow line to drop farther to catch up with the wolves. Although their territory is only about 150 square kilometres (some 60 square miles), one of the smallest among those of the approximately eight packs in the core study area, it is uncanny how an entire pack can disappear.

In times past, the First Peoples by now would have finished salmon processing in their smokehouses and would be in their winter village sites farther into the Coast Mountains. I wondered whether, back then, wolf territories were influenced by tribal groups. Did one pack of wolves follow one tribe of people as they moved from fall salmon streams to the winter big houses? Was the social familiarity between wolves and people that elders in Waglisla describe so strong? Or was the bond diet related? The prey items of people and wolves were similar; both relied heavily on salmon in the fall, deer in the winter, and marine mammals year-round.

Nomads of the Rain Forest

Sometimes, when I look behind me,
I wonder who is observing whom.

TRACKING THE FISH TRAPS

Two weeks later I finally got what I was waiting for. Great big flakes of snow floated down from the platinum sky. Grabbing a pair of snowshoes, I headed towards the boat. At the base of the stairs to our house were three sets of wolf tracks. Members of the Power Line Pack had visited our house sometime in the previous hour, doing the rounds of the neighbourhood and marking everyone's doorsteps—reminding the domestics (dogs) that they don't belong on this island.

I asked William Housty of the Heiltsuk about the relationship between people, wolves and domestic dogs when all the villages were occupied, and he told me of a very old and traditionally secret society, the Dog Eaters, whose story and dance he and the late David Gladstone had recently reintroduced into Heiltsuk culture.

"Our people first domesticated wolves for the purpose of hunting, but it meant taking family members from packs. The wolves in turn would become disappointed and upset that their family was disrupted, and that some of their members left them to become domesticated, that they had betrayed their wild ancestors."

The Dog Eater dance describes how an individual would spend time in isolation before being transformed into a wolf. The dancer would then visit the villages at night and kill the domestic dogs that had left the wild pack, eating parts of them, making them pay for leaving. People would tie red-cedar bark around the dogs as protection from the dancer.

Wolves still visit these outer coastal communities, and each year my neighbours lose a dog or two to wolves.

The tracks I saw now were so recent that I contemplated following them on foot, but this would be my first chance to catch up with the Fish Traps since the salmon run, so I headed out on the boat. The crisp white morning was bloody cold with the wind chill, and although the snow had stopped falling, I could tell by the sky over King Island that the first of the winter outflow winds were beginning.

These winter winds originating from the interior of British Columbia are associated with high-pressure systems. The winds sweep along the glaciers and icefields and funnel down the big channels like Dean, Burke, and Douglas. By the time they hit the outer coast, they can be travelling one hundred kilometres (sixty-two miles) an hour, bringing

Smaller-volume salmon creeks
like this one provide easier fishing
conditions than deeper rivers and
thus provide a disproportionate
benefit for many of the two
hundred wildlife species known
to consume salmon.

with them a wind chill of minus thirty degrees Celsius (minus twenty-two Fahrenheit). These outflows turn ocean spray into ice in seconds and can sink boats from the sheer weight of the ice.

Sometimes, during weather like this, the water collected in my boat's fuel filter freezes during the night. When that happens, I am running on the remaining fuel in the line before the engine shuts down. Melting frozen water in a gasoline line is a tough proposition out here, because using fire is not a good idea. Last time this freeze-up happened, I thought of urinating on the water separator in a desperate attempt to melt it, but in such weather it would merely change the ice to an embarrassing yellow colour.

As I travelled from bay to cove, punching through the thin ice that was forming, searching for signs of the Fish Traps, I wondered how the pups were faring. Since birth, they had known a secure environment revolving around the rendezvous sites, or "headquarters," as some researchers call them. Even the adults, when hunting non-salmon prey during summer and fall, use less than half of their territory, since they prefer to stay close to the pups while also avoiding neighbouring packs.

But come winter, the pack adopts a more nomadic existence, and for the first time since the birthing season the entire family uses the full extent of its territory, often hunting as a unit.

The biggest change is for the pups. Their carefree days of being secure in familiar territory and of being provided for are over. They now travel with the pack, learning how and where to hunt and participating as best they can in kills. They no longer eat regurgitated food or a well-travelled cold leg bone of a deer but taste warm blood and fresh meat.

Winter is also a time of increased competition, as older siblings and parents may be less willing to share. Certainly, provisioning is long over. It is little wonder that winter is a common time for pack members to disperse in hopes of setting up their own territory. How would a human family deal with the sudden addition of five or six offspring all in a single year—especially if the family budget remained the same?

The pups at eight months old are very nearly the size of adults. At ten months most of them, especially the males, are physiologically able to reproduce, but their doing so at this young age, at least in the wild, has seldom been documented. Although many in a pack are of reproductive

Nomads of the Rain Forest

age, typically only one pair mates and gives birth. Under conditions of extremely high food abundance, two or even, rarely, three females in a family may breed, though I have not seen this on the coast.

The alpha female usually employs what is best described as hormonal bullying. As the other females approach estrus, she increases the frequency and severity of her aggressive interactions with them. As a result, their reproductive state is quelled, a phenomenon known as reproductive suppression.

Soon enough the mother-to-be leaves a signal in the snow. When tracking in February, every hundred metres (about three hundred feet) or so I find tiny specks of blood, indicating the female's readiness to breed. Typically the alpha male (but other males too) will mate with her. A remarkably consistent sixty-three days later, a new generation appears.

We humans are not unique in having routines. I knew to start that day's search for the pack in areas where they had had success before.

The snow was so fresh that I could just scan most of the beaches and estuaries from the boat, looking for tracks.

As I finished a survey of previously known kill sites on the east side of the main island, I finally picked up dual sets of tracks about eight kilometres (five miles) to the northwest of the wolves' fishing spot. The tracks crossed a small beach where the pack had taken down a deer the previous summer. Perhaps after that success they had added this place to their scouting repertoire, just as I return each year to places where I have previously seen wildlife.

I went ashore, put on my snowshoes, and began following the tracks. The snow was still light, and I hoped that the cold weather would keep it that way. Following wolf tracks in the snow is totally revealing of what the animals have been doing, and I was excited to have such access to their daily lives. When they depart from the trails to sniff out a squirrel cache or a bear den, I can see that. Nothing is hidden, and I almost feel guilty for invading their lives like this.

It is easy to get used to the predictable summer range, which mostly centres on the pups and the various rendezvous sites established. Now I had to work much harder to keep track of the wolves. The pups were fully mobile, and the deer the pack focusses on in winter were

This is literally a lone wolf, though he would like to change that. Up to 20 per cent of wolves found alone in the landscape are considered dispersers or "extra-territorials"—wolves looking for a pack to join or for vacant territory in which to start a family.

also mobile and widespread; the wolves could be anywhere on close to a dozen islands.

In addition, up to 20 per cent of wolves alone in the landscape are considered dispersers, or "extra-territorials." These two sets of tracks could belong to wolves who had left the pack permanently, or a pair from a different pack altogether, searching for new territory.

Dispersal can reap big rewards if a wolf manages to set up a new territory and breeds, for that means that the genes of that individual will be passed along. Such behaviour is helpful to the population at large, too, since it prevents genetic inbreeding. As a rule, breeding couples choose unrelated mates over relatives.

The strategy is not, however, without risk. Lone dispersers cannot survive as well as wolves that belong to a pack, so it is in their best interest to either join an existing family or find a mate and vacant territory and attempt to start a new pack. If dispersers are not accepted into an existing pack, they will frequently be killed as competition.

Solitary dispersing wolves also suffer a significantly higher rate of human-caused mortality than wolves that stay within family structures. Often such dispersers are

desperate for food and lack the collective wisdom and wariness of a pack.

Even if it survives violent encounters with prey or competing wolves or harsh winters or human hunters, an animal that must kill with its teeth and jaw faces other, less obvious dangers.

Once I saw a young wolf with a porcupine quill lodged in his nose. The quill looked to be buried quite deep. I suspected that it would be difficult for the wolf to hunt—or even eat, for that matter. I also figured that he was avoiding the den site because the playful pups would inadvertently touch the barb and cause added pain.

I thought of pulling the quill out, since he sure seemed to want help, but because he wouldn't be fully constrained I feared getting bitten for my efforts. I had pulled quills out of dogs, and it is painful for the animal. Facing jaw pressure more than seven times greater than humans can exert, I figured that nature needed to take its course. I wondered if the pack would feed this wolf in addition to the pups.

One month later, Paul Paquet and I found the body of the wolf alongside the river. He was only two years old, but he had already succeeded in many hunts, survived two

winters, and avoided deer hoofs, bullets, and other perils that might kill a wolf. In the end, it appeared that he had been done in by a porcupine quill.

IN CONTRAST TO the summer travel pattern, the tracks I was following pulled away from the ocean. I was in for a long day. They took me about a thousand metres above sea level and then turned, traversing the steep side slope through deep snow. The trail followed the contours of the land, and at various intersections deer tracks crossed the wolf tracks. I followed one of the deer trails, curious as to why they were so far above sea level.

It was easier travelling here, because of the tall and fairly consistent canopy, which caught most of the snow. The trees were large and well spaced—a climax forest, the product of many years and many generations of forest succession. The deer, with their stiltlike legs, are not built for snow and prefer this less exposed terrain.

The deer tracks took me to the base of the trees, where long, stringy rafts of green lichen danced in the wind. The deer were up here browsing.

This forest type, distributed in natural patches throughout the landscape, is critical for the maintenance of stable deer numbers throughout the years. It is especially important on the mainland, where snows can be deep and last for many months.

It's easy to underestimate the value of deer winter range in milder winters with less snow. But during so-called catastrophic winters, significant numbers of deer die if they have nowhere to go for shelter.

During snowy winters, larger old-growth stands of forest receive disproportionately high deer use because of the ability of the forest canopy to intercept snow before it accumulates on the ground. John Schoen and Matt Kirchhoff, two Alaskan researchers conducting deer and wolf studies, have shown that it takes only about fifteen centimetres (six inches) of snow to force the deer into these large stands of old forest.

Although climatic conditions for the coast are generally mild, they do vary greatly from year to year. In our area of the mainland, thirty-year averages in snowfall vary from 86 centimetres (39 inches) in Bella Bella to 155 centimetres

(62 inches) at Ocean Falls. Back in the mountains, this figure can easily quadruple.

In southeast Alaska, researchers tally the number of deer deaths after heavy snow years. During those years, weakened and starving deer go to the shorelines to die. There can be so many dead that researchers walk shoreline transects counting corpses.

On Admiralty Island, in southeast Alaska, Schoen and Kirchhoff found that 39 per cent of adult radio-collared deer died during one severe winter. The researchers estimated that the final mortality figure was closer to 60 per cent. Deer populations, and the wolves that depend on them, take a big hit after this sort of catastrophic loss.

One of the reasons so many deer died was that much of the old-growth forest has been logged. Clear-cuts accumulate too much snow for deer. After twenty-five or so years, dense tree canopies form, and they do indeed block snow, but because of the foreign nature of evenly spaced and evenly aged tree plantations, most sunlight is also blocked, leaving the forest floor devoid of food such as huckleberry, salal, and even lichen. Loss of this critical winter habitat for deer is one of the reasons that U.S. federal courts were petitioned in 1993 to list wolves in southeast Alaska under the Endangered Species Act.

No one wants this to happen to B.C.'s coastal wolves. A few years ago, Chris Darimont assembled a team to investigate how much deer winter range was at risk of being lost on the central coast. A large mapping project was established and measured the extent of overlap between forest suitable for deer winter range and forest suitable for logging (the so-called timber harvest land base, or THLB).

The results were compelling. Deer winter range and the THLB covered small proportions of the overall land base— about 10 per cent each. What was of concern, however, was the overlap. Nearly 50 per cent of winter range occurs in the THLB and thus could be targeted for clear-cut logging.

This finding was an important reminder that the effect of forestry on wildlife populations can be disproportionately larger than the percentage of the area affected by logging. When the companies say they are logging only 10 per cent of the forest, that could mean they are logging 60 per cent (or more) of critical habitat for one species.

Even a quick read of scientific reports shatters the myth of deer as "creatures of clear-cuts." In addition to destroying winter range, clear-cut logging creates other problems. The vegetation that appears so abundant in clear-cuts is often of poorer nutritional quality, offering much less protein for deer. This result has been shown in southeast Alaska and confirmed by Raincoast's work. And snow or not, after the plantation-designed forest canopy blocks the sunlight from reaching the forest floor, it effectively eliminates critical deer food from growing. That is why biologists call these close-canopied plantations biological deserts.

AS I CONTINUED up the mountainside, the two sets of tracks turned into many more. I had found the Fish Traps, and it became clear that they were checking out areas where deer browsed on leafless huckleberry bushes; just the little stubs of the branches could be seen. Between the bogs and open areas, everywhere the forest was old and tall and could be providing security and winter browse for deer, I found both deer and wolf tracks. Soon the wolf tracks formed a single line and dropped back down to an adjacent valley, moving inland to one of the large lakes that formed the core of the island.

The more I travelled through these vast stands of ancient cedars and hemlock, the clearer the problems in wolf-deer dynamics in logged areas became. When small patches of old-growth islands are left in a sea of clear-cuts, these patches become the only suitable habitat for deer during snows. Using the roads associated with logging, wolves can quickly and easily surround and target the deer holed up in these little clumps of forest. In this scenario, wolves can wipe out a lot of the deer in a short period of time. So for that winter the wolves enjoy an abundance of deer, but in subsequent years the deer population crashes and inevitably the wolf population also plummets.

It took the whole day's available light for me to find out where the Fish Traps had been. If the wolf research project used the more invasive technique of radio- or satellite-based telemetry, I most likely would have located the wolves sooner. But then, I reflected, how would that information change our view and understanding of these wolves? Would the data from a more invasive study warrant the information gleaned?

It is late fall, and this young wolf will join the pack in the hunt, for the first time fully exploring the extent of his rain forest home.

Once again, the answer was no. I basically knew where the wolves were (up there on the slope somewhere), why they were there, and what they were doing. Knowing the details was more about my curiosity than about assisting the wolves. And knowing where they were not told me plenty, too.

STUDYING THE HOME RANGES

While the genetics lab at the University of California slowly and methodically processed the scat-based wolf DNA, Chris and I had many conversations about the size of the home range of the various packs within the study area. We anxiously awaited the results of the lab work so that we would be able to test our estimates of which packs claimed which islands.

This lab work is still unfolding. The researchers are identifying the territories of individuals and of all the packs, not just the Fish Trap Pack, within the core study region. From this work, by extrapolation the first DNA-based population estimate of wolves up and down B.C.'s coast will be possible.

Years ago I found it hard to believe that a pack I observed one morning on the north end of an island could be the same pack I saw on the other side of the island that

evening. But now few things surprise me about wolves' ability to travel long distances in short periods of time. They can easily journey from one side of an island to the other, traversing ten kilometres over fairly tough ground in about two hours, and they have been clocked at speeds of more than fifty-six kilometres an hour.

Paul Paquet told me about a particularly peripatetic female wolf in the Rockies, named Pluei. Her family had been strangled in snares and she was on her own, but she was wearing a radio collar so that the researchers could follow her; it transmitted to a satellite receiver rather than to a land-based receiver. They knew that she had been born in the Bow Valley near Lake Louise in Banff National Park and belonged to the Bow Valley Pack, but she soon joined the Banff Pack.

"Based on the signals we were suddenly getting from the satellite, it appeared that Pluei had just jumped into the back of a pickup truck," said Paul, senior adviser for the project at the time. It turned out that she had decided to go for a very, very long walk and was determinedly heading due south.

"We eventually tracked her into Waterton National Park and across the U.S. border all the way down to Yellowstone,

Nomads of the Rain Forest

where she abruptly turned north again and headed back up into the Elk Valley in southeastern B.C., making her way all the way back to Banff," Paul said.

She lived for another three years before ending up dead, shot by a hunter near Invermere. It turned out that she was one of the wolves Chris had followed during his time in the Rockies. But so many wolves that he tracked ended up dead. In fact, the amount of human-caused mortality was one of the reasons the project shut its doors. "We literally ran out of wolves to study," lamented Paul.

One of the difficulties in estimating home range size is that natural boundaries sometimes have little influence on territorial area—there is no surefire way to identify a territorial boundary. Wolves swim across broad stretches of water; we now know that they can swim more than ten kilometres of open ocean. On the mainland they do not avoid the high country but some groups likely claim it as an important prey base for mountain goats and possibly marmots.

Nevertheless, I felt that the information gathered from radio collaring did not justify the intrusion on the wolves. In addition, flying in bush planes or helicopters is dangerous. According to Chris, "Right from the start, when we

took stock of our resources, we realized that we were short on cash but rich with talented volunteer field workers, local knowledge, and time. I knew we could do this research another way."

Many of the research papers on coastal grizzly bears are telemetry based, and to see the polygons plotted of the bears' seasonal movements is fascinating. In the case of the Khutzeymateen River estuary and valley—part of which form Canada's first grizzly bear sanctuary—the technique might have helped lead to the bears' protection. On reflection, however, I think that a noninvasive study would have been equally informative in the Khutzeymateen, if not more so. All that money spent on helicopters and planes could have gone to field researchers, and when you have people in the field, gathering data by direct observation, important information is gleaned that no transmitter on an animal's neck can duplicate. And regrettably, some bears died for the cause, killed in the snares used to set the collars on them. It doesn't get any more invasive than that.

The other major study of grizzly bears in the Great Bear Rainforest—in the Kimsquit Valley—was also invasive. Again, some bears died, and for several years others

A grizzly bear takes a nap between salmon meals, but it isn't necessarily safe in doing so. British Columbia still sanctions trophy hunting of coastal grizzlies even in most protected areas or parks, and most of the slaughter comes on the estuaries in spring and the salmon rivers in fall.

carried the burden of collars that weighed up to one and a half kilograms (about three pounds). After all was said and done, the sponsoring company, Western Forest Products, ended up ignoring all the findings, cut down pretty much every tree in the valley, then shut down operations in the early nineties. What was left was a seriously degraded valley, landslides, biologically impoverished tree plantations, and a road network for hunting and poaching more bears. The harassment of the Kimsquit bears didn't help their cause or the once-great salmon runs.

I cannot imagine doing the same to these wolves. First the wolf has to be trapped and tranquillized—aerial gunning is not possible in the temperate rain forests. The wolf is snared by a steel cable in a baited trap (accounts of wolves chewing off a leg to free themselves from leghold traps are not uncommon), and then the biologist and crew approach the wolf and tranquillize it with a drug. The drug only immobilizes the animal; it does not render it unconscious. The wolf observes everything that happens to it, and for the first time in its life, the apex predator of the coast can do nothing to help itself.

The anaesthetic ketamine, or "Special K" as it is known on the streets, is one of the suite of drugs used, and it leads to dreamlike dissociative states and hallucinations. When used as an anaesthetic in humans, it is combined with another drug to prevent hallucinations. Wolves are not as lucky. It is difficult to regulate a "dose" of ketamine, and there is only a slight difference between the desired effects and an overdose. It's a strong depressant at higher doses and can dangerously reduce heart rate and respiratory function.

Imagine what alpha males or females feel when this is happening to them, how they must suffer as leaders for failing the rest of the pack. And imagine how the rest of the pack must feel to have to abandon the trapped individual at the last moment as the biologists approach. In terror, the wolves hide in the dark forest, listening, smelling the fear of their packmate as he or she is "processed."

The "process" usually involves pulling a tooth to properly determine the age of the wolf. Its gums are tattooed for future identification, a collar is buckled tightly around its neck, a blood sample is taken, and the body is hoisted on a scale to measure its weight. Once all this is done, the

A coastal grizzly bear looks for
those last few meals of salmon
before heading into the mountains
to hibernate for the winter.

Nomads of the Rain Forest

The Fish Trap Pack takes a break from salmon fishing. Standing in the centre of the image and looking otherwise fit and healthy is the pup-sitter I called Three Legs, or TL, her permanently disabled hind leg visible.

Nomads of the Rain Forest

biologists pack their gear and move away as quietly as they can and leave the wolf to regain its mobility as the drug wears off.

Then the researchers begin to track the wolf by flying overhead in a small airplane or, if they can afford it, a helicopter. They fully expect the wolf to go on with its life as naturally and normally as on the day before it became a research subject.

I have found that biologists tend to jump at the chance to put collars around an animal's neck without first asking a simple but critical question: Will this information change the way we humans behave around wolves? After all, ultimately this behaviour is not a wolf problem but a human problem. If they do ask the question, many times the answer is no, but the study goes ahead anyhow: "Well, we have some funding."

I can see that radio collaring is tempting because it involves lots of gadgets and computers, and technology must appeal to a scientific mind. But I think, finally, it is perceived as simply the easiest way. It is infinitely easier than what Chris and his team are doing. But is it as effec-

tive, and is "easy" a justification for the harassment of such an intelligent and social animal?

I GOT LUCKY. My neighbour, who is a handlogger, told me that the day before, while he was beachcombing for logs, he had seen some wolves on an islet in the Fish Traps' area, and it looked as if they were eating a sea lion.

Going in slowly, I glassed the bay with the binoculars. I first located the ravens and then focussed on two wolves sleeping on their sides. If they had not been feeding recently, they would have been sleeping curled up tail to head, in a ball. But with full stomachs, they were stretched out next to the remains of what was indeed a sea lion carcass. I set up a telescope about a kilometre across the channel and waited.

None of the wolves were feeding on the carcass when I arrived, but I could see flesh and bones strewn about the beach. A flock of ravens landed on the rocks, and within seconds two other wolves raced out of the forest and chased them off.

A breakthrough in the understanding of wolf-raven relationships came recently in Yellowstone National Park.

The white around this wolf's muzzle
indicates that he is an older wolf.

Researchers measured the amount of each wolf-killed car-cass that ravens scavenged. Every bite to a raven is one less to the packs. It was found that individual wolves and small packs lost disproportionately more food to ravens than did larger packs. Given that individual wolves can easily kill large prey, the researchers hypothesized that it was losses to ravens, not the necessity to hunt communally, that was responsible for the evolution of pack formation in wolves.

The two wolves were TL and the white female, Urchin. Soon the rest of the pack came out of the forest, and I could see the steam rising from their breath as they buried their heads in the cavity of the huge sea mammal. Their muz-zles were soaked in blood. Three of the pups were playing tug-of-war with about fifteen metres of intestine.

Seeing TL was good news. She and the other wolves had swum across two bodies of icy water and paddled across the channel to reach this islet, and I liked to see her hav-ing a decent and long feed. This kill was an important win-ter bonus of marine nutrients for the Fish Trap Pack, the equivalent of more than half a dozen deer, and given the size of the sea lion it was most likely scavenged, so they had expended little energy in catching it.

The day was going fast, and I was getting cold. The pack had finished eating and moved off the kill, and I thought about heading home. But a movement above the tree line caught my eye; I aimed my telescope towards a granite out-cropping rising out of the centre of the island, about 150 metres above sea level. There was White Cheeks, emerging from the forest. A few moments later the rest of the pack followed. From their vantage point, they could survey the full western edge of their territory.

I thought that they must be pleased with their situ-ation. All of this year's pups were still alive and healthy. I picked out Ernest, fully grown and looking as beautiful as ever, staring out over the water. Even from this distance, it seemed as if he were up to his old tricks and staring at me. As I focussed on him with the telescope, I whispered a greeting.

White Cheeks, standing on the edge of the cliff, arched his back and began to howl. The rest of the pack, with full coats and full stomachs, moved to the edge and joined in. It was one of those winter days when the air is so still and so quiet that sound seems to amplify with distance. The howls picked up an ecstatic momentum. As they cried, some of

Nomads of the Rain Forest

the wolves stood on their hind legs, pawing at the air with heads held back, as if to loft their voices higher into the sky.

As quickly as the howling had started, White Cheeks silenced the pack. For a few seconds, the wolves licked and nuzzled each other as their song rolled over the rain forest, undulating across the ocean and lakes, over the weathered cedar haunts of the old village sites.

I closed my eyes, savouring the sound. I thought of the countless animals, many of them prey, alerted into stillness as the wolves' celebratory song floated past.

The pups resumed playing, until two minutes later I heard howls again, apparently returning like the incoming tide. I thought at first it was the longest echo ever, but then I realized the impossibility of that. Every member of the pack became instantly attentive and stared off to the north. Faint sounds, fading in and out, but louder now—it was an answering howl. The Village Pack was responding.

The emotion that swept over me reminded me of how I had felt listening to the old songs in the big house, the drummers beating ancient rhythms on the hollowed-out cedar logs; that was a privilege that I didn't take lightly. I felt the same way as I sat here between two wolf families whose ancestors have walked these forests for so many thousands of years. I wondered what they had said to each other. And I wondered how the Village Pack was faring this winter. I thought of the Surf Pack and the storms they would still face before spring arrived.

I watched as, one by one, without a backward glance, the Fish Traps faded into the forest. I looked around and just felt silence and a loneliness. I shuddered in the cold air.

It had been a good year, like all the rest.

Canada's rain forest
wolf—the eyes of an
animal yet unbroken.

EPILOGUE

The Eyes *of the* Wolf

. . .

I HAVE BEEN blessed to observe wild wolves in a landscape of unprecedented beauty and grandeur. But now there are river valleys and islands to which I can no longer return, places intact when I knew them but now brutally and efficiently altered.

The wolves of James Creek on Pooley Island were among the first to teach me about wolf society, and my greatest memories are of quiet summer evenings spent sitting among the pack in the fields of dune grass, watching them herd salmon into the shallows, or lying wide awake in my bunk and listening to their howls as they hunted in the night. These are bittersweet memories.

Years of struggle to protect Pooley Island ended the day I watched the first barges loaded with road-building equipment land in the bay. The first dynamite blast ripped the main wolf trail in half and sent pieces of it flying across the water. I left that day, turning my back on the wolves that had taught me so much, never to return.

Sailing past Pooley Island, now covered in a matrix of roads and clear-cuts, I often think of what is left of that once-proud pack. I think of them at night, shielded by the black shadows, surrounded by the stench of the work trailers, the generators, and the garbage from the logging camps. I think of the reek of diesel oil, the smell of the domestic dogs and the potential diseases they bring, the dynamite cord that Chris Darimont identified in the wolf scat samples, the guns and trucks. Roads now criss-cross the islands, where once only old, thin, well-worn trails skirted the heights of land and revealed where the deer slept. Now the wolves walk the roads at night. Their eyes are still bright, but they have reason to reflect fear and uncertainty as their habitat is destroyed.

A few of my neighbours, both Native and non-Native, have been employed in the logging industry on Pooley and other places on the coast. Even they admit that the scale and rate of logging are unsustainable. The logs are still shipped down south or offshore, so local workers are restricted to cutting the trees and building roads. At the current pace, it will be only a matter of years before the ecological and economic collapse of the rain forests to the south and on Vancouver Island is duplicated here.

In February 2006, the British Columbia government announced new conservancy designations for approximately 30 per cent of the Great Bear Rainforest. In addition, industry, First Nations, and the provincial government made a commitment to implement ecosystem-based management (EBM) practices on the rest of the land base by 2009. Although this announcement was heralded as a "rain forest victory," it is uncertain what EBM will mean for the 70 per cent of the coast that is unprotected. For this reason, a declaration of victory seems premature.

Wolves were not even considered in designing and choosing protected areas. As a result, not a single such area is likely large enough to "protect" even one pack's full territory. Furthermore, the agreement does not ensure the preservation of travel corridors between protected zones, a necessity in safeguarding far-ranging large carnivores like bears and wolves.

Given the multiple-use objectives within the conservancies, it is difficult to consider them truly protected. For example, mining allowances were made for approximately 5 per cent of the conservancies. Fixed-roof accommodation for large-scale tourism ventures is being promoted within some of them. Most troubling, trophy hunting of wolves and bears is allowed in most of the new conservancies.

Proponents of the plan believe that EBM will compensate for the lack of core habitat protection. But this model has not proved successful in other jurisdictions, and no concrete evidence exists that it will conserve the Great Bear Rainforest and the species that depend on it over time. Right after the Great Bear Rainforest agreement was announced, clear-cut logging and road building escalated in many intact rain forest areas, offering little reassurance that industry and government were committed to changing forest practices.

In addition, even if forest practices improve in the areas that adopt an EBM model, logging roads will still be necessary to access intact rain forest. In his exhaustive analysis of wolf mortality in southeast Alaska, Dave Person found that thirty-nine out of the fifty-five wolves that were being studied between 1993 and 2004 died. An incredible 82 per cent of these deaths were directly attributed to hunting or trapping. Seventy-five per cent of the mortalities were caused by people accessing the wolves from roads.

Because of this artificial and very high level of human-caused mortality over the past one hundred years, the wolves to the north have likely been pushed through a genetic bottleneck. Emerging results from collaborators at UCLA and Sweden suggest that these Alaskan wolves have lost the genetic diversity found in coastal B.C. wolves like the Fish Traps and their kin. It is this diversity that makes British Columbia wolves so special and among the truly last wild wolves on the planet.

In 2005, in an unprecedented step, the Raincoast Conservation Foundation, with the support of the central coast First Nations, purchased the licence for one of the largest guide-outfitting territories in B.C. The licence covers almost half of the Great Bear Rainforest area and applies to the territories of the Fish Trap and Village packs, among others. It includes exclusive commercial trophy-hunting rights and thus will further protect wolves and bears in at least part of the rain forest from commercial trophy hunters.

Clear-cuts denude the Parker
Creek region of King Island in Dean
Channel, October 2006. Shortly
after the Great Bear Rainforest
agreement was announced in
February 2006, large-scale clear-
cut logging escalated in many
coastal watersheds.

FAR RIGHT: Hawkesbury Island in
Douglas Channel, October 2006.

As chief councillor Ross Wilson of the Heiltsuk Nation said at the time the purchase was announced, "We hunt out of need, not desire," a sentiment I hear from the leaders of First Nations up and down the coast. To the north and south of Raincoast's licensed territory, however, commercial trophy hunting continues. And B.C. resident hunters can still kill up to three wolves per person per year with no special licence.

Sport hunting of large carnivores like grizzly bears and wolves should be banned, a step that has the support of most of the Canadian public and would strengthen a burgeoning wildlife-based tourism industry.

Although some progress has been made in protecting part of the rain forest, virtually nothing has been done to protect the marine environment. Studies have shown that up to 80 per cent of marine-based nitrogen fertilizing the riparian forests comes from the remains of spawning salmon. In addition, Chris's study shows that 50 to 75 per cent of the diet of outer coastal wolves consists of marine-based prey, and salmon figure prominently. Yet the commercial fishing industry can still take up to 80 per cent of some salmon runs, leaving only a minimum "escapement target" to meet management quotas for salmon reproduction.

Another significant threat to the survival of wild salmon is the effect of farmed salmon raised in open-net cages. Approximately one hundred salmon farms now exist to the south of the Great Bear Rainforest. Biologists are documenting the loss of wild salmon to parasites, and they have great concern about the spread and amplification of disease and pollution associated with the farms. Whereas European countries are finding that the way to protect wild salmon is to separate them from farm salmon, Canadian management is risking more than one-third of B.C.'s wild stock by densely situating farms on the west and east coasts of Vancouver Island.

As well, Atlantic salmon, the favoured species to farm in Pacific waters, are escaping, surviving, and reproducing. These aliens have now been found as far north as the Bering Sea—thousands of kilometres from the nearest fish farm. Alaska has banned salmon farming, but the B.C. government continues to push for expansion of the farms in the Great Bear Rainforest.

Epilogue

In Canada, the federal government has jurisdiction over tidal waters. In 1972, Ottawa placed a moratorium on oil and gas exploration and tanker traffic on the B.C. coast; remarkably, considering the world's thirst for oil, this moratorium has remained in place. But the lure of oil can be kept at bay only for so long. In 2005, B.C. premier Gordon Campbell presumptuously announced that oil from the province's new offshore reserves would be used to light the Olympic torch for the 2010 Winter Olympics to be held in Vancouver, even though the province does not have jurisdiction over marine resources.

Currently, pipelines to Kitimat and Prince Rupert on the north coast are being planned. If they are built and the moratorium is lifted, tankers approaching B.C.—carrying condensate used to liquefy crude in the pipeline—will pass outbound tankers exporting crude oil to be shipped to Asia and other world markets. These tankers will travel through the heart of the Surf Pack's territory, as well as the territories of many other packs. We know from experience in other offshore oil-producing regions that it is not a matter of if but when a spill will happen. Nearly twenty years later,

researchers are still calculating the environmental damage caused by the *Exxon Valdez* oil tanker spill in Alaska. When one of these tankers slams into any one of the thousands of islets and islands that form the backbone of the Great Bear archipelago, wolves will be just one of countless casualties.

Time and time again, I have been taken aback by the extraordinary ability of wolves to assess situations, especially danger and threats from great distances. I frequently have to remind myself not to treat photography as a "pursuit" of the wolves. What I feel while following them is too similar to what a hunter feels if I view it as pursuit, and wolves easily sense this. They tend to appear only when I have convinced myself that it doesn't matter whether they show up that day or not.

Another example of wolves' extraordinary instincts occurred in 2006. Gudrun Pflueger took time off from the wolf research project to guide a German television film crew working on a documentary on coastal wolves. It was a dream job for Gudrun; as a change from running line-transects and picking up deer and wolf scat, she spent long hours searching for wolves for the film crew.

The howls are not being heard: so far the needs of wolves, the apex predator of the coastal rain forest, have not been considered in developing long-term conservation plans for the coast of British Columbia.

Later in the summer and at the end of the film project, Gudrun was lying in the middle of an estuary, the camera operator a good distance away, when the most extraordinary event occurred. As she lay in the grass, a pack came out of the trees and surrounded her. It was peculiar behaviour for a full pack to get so close so quickly, at the first encounter, especially with more than one person in the area. The wolves surrounded Gudrun, smelled her, walked around her, and watched her from only a few metres away. They were not especially upset and did not show aggression, which would have been more typical of such an intrusion. They spent the rest of the evening playing nearby, almost as if they did not want to leave her. Marven Robinson of the Gitga'at First Nation, who was also working on the film and watched the episode unfold, noted that something was very different about how the wolves reacted.

Two weeks later Gudrun was in hospital, having just been diagnosed with a life-threatening brain tumour. At this writing she is still fighting for her life. When I watch the footage from that evening, I can't help but think that the wolves knew that she was ill and were trying to communicate something. Chester Starr often talks about a Heiltsuk belief that wolves do not show themselves unless they are trying to tell us something.

IT CAN BE difficult to balance the needs of wolves and the need to educate the public about wolves. People are not interested in protecting what they do not know and understand. But exposing wolves to too many people has not been without consequence.

In 2004, Raincoast helped a crew from National Geographic Television set up to film a pack whose territory was next to the Fish Traps' domain. The crew settled in for a long season, and soon they were getting results. Over the months, they filmed the wolves interacting with each other, with grizzly bears, and with black bears, as well as their exceptional salmon-fishing behaviour. The team knew each wolf and watched each of the pups catch its first salmon. The footage was remarkable and would soon be a film seen throughout the world.

By the end of October, the salmon had finished spawning and the film crew was gone. In November I received

Epilogue

a letter from the local commercial guide-outfitter, who at that time lived in Bella Coola. He informed me that "our" precious wolves existed no longer, that he had just killed as many of them as he could; he called it "ungulate enhancement." He had shot them dead as they played on the beach. I found out later that he had discovered where the National Geographic crew had been filming. Raincoast purchased the outfitting licence a year later.

I often think that if the wolves had not been conditioned to the film crew that season, they would not have shown themselves to strangers so willingly. How ironic that those wolves, having escaped the hand of man for so long, would be dead within the same season during which they would be seen in the living rooms of millions of people around the world.

IT IS NOT just the commercial trophy hunters who target wolves. In 2001, Chris and I ran into a couple who had recently moved from Idaho to the remote abandoned mining town of Alice Arm on the north coast. When we pulled up to the dock, immediately doubling the town's popula-

tion, Chris asked if the couple knew where to locate wolves so that he could search for scat. The man said he could get us something better than scat.

He took Chris out into the inlet in his skiff. After pulling them up at the end of about sixty metres (two hundred feet) of rope, he dropped his prawn traps into the bottom of the skiff. Inside each trap, prawns crawled through the eye sockets of a partially eaten wolf skull.

Some people have so little respect for wolves that they kill them for prawn bait.

A WEEK LATER, on the way home, we stopped in at the abandoned cannery at Butedale, where the caretaker told us we could find a dead wolf in the blackberry bushes behind the house. He had just shot it, because he feared that the wolf would kill his dogs. He further informed us that the best way to catch wolves is to hide large halibut hooks in a hunk of meat and hang it from the branch of a tree about two metres above the ground. When the wolves jump up for the meat, they are caught by the hooks or swallow the hooks whole, ripping out their insides.

Over the last three hundred years, human beings have reduced global wolf populations by 80 per cent, and the wolf's former range has decreased by 40 per cent.

THESE ARE JUST a few stories that show what wolves still have to contend with.

In 2001, having just found the den site under the fallen cross timbers of the Heiltsuk big house on Yeo Island, we heard dynamite blasting and large machinery making its way across the upper slopes about five kilometres (three miles) to the south. Western Forest Products was building a logging road going right towards the den site. Chester and I hiked up the slopes above the village and the den site and found flagging tape outlining all of the cut-blocks. Within one cut-block, we found a very old canoe-manufacturing site and counted culturally modified trees in the hundreds.

On the day we showed Heiltsuk representatives the village site to which the wolves had led us, and the proposed logging site, the pups came rolling out of the forest as if on cue and played on the beach for several minutes.

Shortly thereafter, the Heiltsuk leadership was able to persuade the logging company to stop the road construction.

But before the company pulled out, one of its trucks accidentally ran over the alpha female of the Village Pack and killed her. The driver threw the dead wolf off the side of a bridge. The rest of the pack stayed on for three days, howling, mourning the loss of its highest-ranking female, the mother of that year's pups. Yet it has been said that one thing that separates humans from other creatures is our ability to mourn the loss of our dead.

Today the unfinished road sits pointing at the den site. The Village Pack's territory remains unprotected, as does the Fish Traps' sector. The Surf Pack has received partial protection under the Great Bear Rainforest agreement.

When I stare into the amber eyes of a wolf, I feel the closest to understanding an animal whose blood flows with the confidence of one that has never been broken. Those are the eyes of a hunter who has never been hunted. Those eyes offer a portal into understanding not just wolves, but also the rain forest world they represent. When I look into those eyes, I ask for a bit more time. I ask the wolf to be patient with us a little longer while we find our way.

Future generations may very well judge our success by how bright those eyes still shine.

Acknowledgements

MANY PEOPLE have contributed to this book in countless ways, and I am deeply grateful to you all. To those that are not mentioned below, I hope that I have thanked you in other ways.

In particular, I am blessed with my wife and soulmate, Karen; my daughter, Lucy; and my son, Callum. I thank my mother, Jane, for her support and encouragement, and my father, Peter, who first saw the conservation opportunities in the Great Bear Rainforest and acted accordingly.

Thanks also go to the following people:

To Cameron Young, for his long-standing commitment to B.C.'s rain forest and his patience and guidance, which helped form the early stages of this book.

To everyone at Greystone Books, in particular Rob Sanders for supporting this project from day one and Nancy Flight for her supportive guidance and wordsmithing, and to Wendy Fitzgibbons for her cheerful final copyediting.

To the staff, directors, volunteers, and supporters who have enabled Raincoast to make valuable contributions to conservation on the B.C. coast. I especially want to thank Chris Genovali, Misty MacDuffee, Nicola Temple, Chris Darimont, Heather Recker, Robin Husband, Jennifer Kingsley, Michelle Larstone, Mike Price, Rob Williams, Chris Williamson, Corey Pete, Marnie Phillips, Teunis Jan Schouten, Loredana Loy, Shelby Temple, Will Cox, Frances Hunter, and Briony Penn.

To the captains and crew members, whose perspectives and observations I value greatly: Erin Nyhan and Brian Falconer, *Achiever*; Trish Smyth and Eric Boyum, *Great Bear 11*; Kevin Smith and Maureen Gordon, *Maple Leaf*; Jenn Broom and Tom Ellison, *Ocean Light 11*; Doug and Carol Stewart, *Surfbird*; Jean-Marc Leguerrier, *Til Sup*; Stan Hutchings and Karen Hansen, *Hawk Bay*; Dave and Stacey Lutz, *Nawalak*; Harvey Humchitt and Mel Innes, *Clea Rose*; Mike Durban, *Blue Fjord*; Randy Burke, *Island Roamer*; Patrick and Marsha Freeney, *Nirvana*; Vern Sampson, *Frances M*; Ralph Nelson, *Gnoses*; Warren and Helen Buck, *Metridium*; Mike Hobis, *Duen*; Gary Housty, *Twin Fisher*; Robbie and Jan Macfarlane, *Merry Mac*; Rob Flemming *Pender Chief*; Larry Olsen and Dave Bell, *Canadian Shore*.

I have drawn a great deal from the following writers, scientists, researchers, and biologists over the years, and I am forever grateful. Wayne McCrory, Paul Paquet, Michael Soulé, Richard Jeo, Sanjayan Muttulingam, Alexandra Morton, Barrie Gilbert, Lance Craighead, Matt Kirchhoff, John Schoen, Dave Person, Dennis Sizemore, Barry Lopez, Brian Horejsi, Bristol Foster, Faisal Moola, Morgan Hocking, Dan Klinka, Neville Winchester, Dionys de Leeuw, David Suzuki, Rick Bass, L. David Mech, Doug and Andrea Peacock, Peter Ross, Ian McTaggart-Cowan, Chris Filardi, and Tom Reimchen.

For assistance with photography I thank Chris Cheadle, Jeffrey Bosdat, Adrian Dorst, and Marvin Nehring, and Custom Colour for superb processing.

Although any errors or oversights lay firmly with me, I thank the following for reviewing parts or all of the manuscript: Karsten Heuer, Leanne Alison, Chris Genovali, Wayne McCrory, Chris Darimont, Paul Paquet, and Jane McAllister.

To the many friends and colleagues who have contributed to this collective work over the years: Bryan McGill, Don Arney, Twyla Roscovich, Ellie and Kiff Archer, Evan Loveless, Johanna Gordon-Walker, Gudrun Pflueger, Murray Reid, Fred Reid, Andrew Kotaska, Christine Scott, Andrew Westoll, Juergen Boden, Sam Catron, Jennifer Carpenter, Mary Vickers, Don Vickers, Chuck and Phoebe Rumsey, Sam Tucker, Martin Campbell, and Ed Moody. The late Ed Martin, J.R. Martin, David Gladstone, and Cyril Carpenter. Ross Wilson, Jordan Wilson, Elroy White, Nicholas Read, Brian Payton, Matt Jackson, Liz and Ron Keeshan, Svetlana and Jeff Hansen, Doug Neasloss, Marven Robinson, Chester Starr and Pic Walker, Uwe Mummenhoff, Michael Mayzel, John Huguenard, Charlene Wendt, Mike and Maureen Heffring, Ian Gill, Baden Cross, Stephen Anstee, Heidi Krajewski, Anita Rocamora, Sandy and Savvy Sanders, Jessie Housty, Marge Housty, and Larry Jorgenson.

I am especially grateful to William Housty for his patience and valuable insights into Heiltsuk culture and for explaining and sharing the Dog Eater Society story. I also thank T'sumklaqs, Peggy Housty, who kindly gave permission for the wolf story to be included in this book as well as images of her wolf mask and hides, and to Pauline Waterfall for guidance.

Finally, Chris Darimont and I thank the supporters of Raincoast's wolf research, in particular the Wilburforce Foundation, the Vancouver Foundation, Raincoast Conservation Foundation, Mountain Equipment Co-op, Patagonia, Inc., Robert and Birgit Bateman, the McCaw Foundation, the National Geographic Society, the Summerlee Foundation, the Valhalla Wilderness Society, the University of California—Los Angeles, the University of Victoria, Natural Sciences and Engineering Research Council, the Bullitt Foundation, Susan Mackey-Jamieson, Yvon and Malinda Chouinard, the Heiltsuk Hemas and Tribal Council, Nathan DeBruyn, Song Neo-Liang, Bo Reid, Shelley Alexander, Merav Ben-David, Heather Bryan, Jennifer Leonard, Erin Navid, Rick Page, Mike Quinn, Gordie Gladstone, Kasia Rozalska, Patty Swan, Robert Wayne, Michael Uehara and King Pacific Lodge, Dean and Kathy Wyatt, Knight Inlet Lodge, Craig Widsten and Shearwater Marine, and Jim and Jean Allan for providing a warm shelter during the writing of this manuscript.

For more information, contact me at PO Box 26, Denny Island, BC, Canada VOT 1BO, or at www.ianmccallister.org.

PHOTOGRAPHY NOTES

All of the wildlife images in this book are taken in natural and wild circumstances. No image has been altered or manipulated. Film stock was mostly 35mm film using Fuji Velvia with occasional Fuji 100. Nikon and Pentax equipment was used. I am grateful to LowePro, Nikon, Patagonia, and Raincoast for equipment donations and to Lens and Shutter and Vistek for technical support.

Acknowledgements

Index

Italicized page numbers
indicate figure captions.

bears. *See* black bears;
    grizzly bears
beavers, 102
Bella Bella (Waglisla), 9, 27, 34
black bears, 16, 20, 50, 63–70.
    See also grizzly bears

Campbell, Gordon, 183
coastal wolves: mainland vs. island
    populations, 49–50; relation-
    ship with humans, 59–60, 70,
    112, 132, 148, 151, 183–84; sub-
    species of gray wolf, 7–8, 23–24,
    53–54. *See also* Fish Trap Pack;
    gray wolves; Power Line Pack;
    Surf Pack; Village Pack
conservation issues, 6–8, 160,
    163, 168, 179–87

Darimont, Chris, 11–12, 29, 145,
    167; deer winter range study,
    160; dietary analysis, 63, 70,
    110, 141, 144, 180; research style,
    47–49, 54–55. *See also* research
    methods
deer, 59, 69; critical winter
    habitat, 159–60, 163; popula-
    tion surveys, 56; as wolf prey, 66,
    102, 119, 128, 131, 155–56

den sites, 37, 40, 40, 43–44, 55–56,
    59–60, 123–24, 127
diet, 8, 49, 105–6, 110; bears in,
    63–70; berries in, 108; deer in,
    66, 102, 119, 128, 131, 155–56;
    marine resources in, 105, 106,
    108, 109, 138, 144–45; salmon
    in, 23, 128, 131–32, 134, 137–38,
    141, 145; seals and sea lions in,
    105, 110, 173–74
dispersers, 156

ecosystem-based management, 180
Enns, Maureen, 78

First Nations, and coastal wolves,
    24, 26, 59–60, 70, 112, 132,
    148, 181–82. *See also* Heiltsuk
    First Nation
fisheries management and fish
    farming, 141–42, 144, 182
Fish Trap Pack, 27, 39, 44, 142, 155,
    171; conservation status, 187;
    den sites, 37, 40, 43–44, 123–
    24; diet, 54, 131, 132, 137–38,
    144, 173–74; division of labour,
    43–44; dominance over grizzlies,
    16, 19–20; "Ernest," 37, 124,
    127–28, 147; genetic and DNA
    analyses, 50, 53–55; home range
    and movements, 44, 47, 89–90,
    163–64;

howling, 19, 174, 177; injuries
from prey defences, 43–44;
pups and pup behaviour, 37, 40,
43–44, 47, 73, 123, 124; ravens
and, 34, 37, 142; reproductive
behaviour, 40, 43–44; scat
analysis, 50, 53–54; "Sentries,"
40, 132, 135; social hierarchy and
bonding, 19; swimming, 44, 47,
47, 120; "Three Legs," 43–44,
144, 171, 174; "Urchin," 37, 174;
"White Cheeks," 37, 38, 137, 174;
winter survival, 174, 177
forest industry, 8, 160, 163, 168,
179–80, 181, 182, 187

genetic and DNA analyses, 29,
49–50, 53–55, 164, 181
glaciation, and coastal fauna distribution, 50
Gladstone, David, 151
gray wolves: as charismatic species,
8, 11; coastal vs. mainland subspecies, 8, 53–54; distribution
and abundance, 6–8, 186.
See also coastal wolves
Great Bear Rainforest, 8, 9, 90, 148,
173; conservation legislation,
180–83, 187; homesteaders in,
103, 105; intertidal habitats, 15,
70, 81, 123; marine environment,
threats to, 8, 182–83; Raincoast

Archipelago, 74, 74, 77; weather
patterns, 151–52, 159–60.
See also forest industry
great blue herons, as wolf prey, 105
grizzly bears, 16, 19–20, 23–24, 30,
138, 167–68, 170. *See also*
black bears

habitat protection. *See*
conservation issues
Heiltsuk First Nation: conservation efforts by, 187; dances and
myths, 27, 60, 115, 151; fish traps,
40; the "Gateway," 47; relationship with coastal wolves, 26, 44,
47, 56, 59, 151, 184; traditional
ecological knowledge, 29,
47–49; Waglisla, 9, 27, 34
home range and movements, 20,
152, 164, 167, 180; Fish Trap
Pack, 44, 47, 89–90, 163–64;
Surf Pack, 78, 89–90, 102–3,
105–6, 112–13, 116
howling, 19, 101, 174, 177
human-caused mortality: of bears,
167–68; of wolves, 6–7, 128, 131,
160, 167, 181, 182, 186–87
hunting and trapping. *See* human-
caused mortality

injuries from prey defences, 43–44,
156, 159

191 }